THE BIOLOGY OF PLANT-INSECT INTERACTIONS

A Compendium for the Plant Biotechnologist

THE BIOLOGY OF PLANT-INSECT INTERACTIONS

A Compendium for the Plant Biotechnologist

Editor

Chandrakanth Emani

Department of Biology
Western Kentucky University-Owensboro
Owensboro, KY
USA

CRC Press is an imprint of the
Taylor & Francis Group, an **informa** business
A SCIENCE PUBLISHERS BOOK

Cover credit: Figures 1 and 2 from Chapter 5. Reproduced with kind courtesy of the authors of the chapter.

CRC Press
Taylor & Francis Group
6000 Broken Sound Parkway NW, Suite 300
Boca Raton, FL 33487-2742

Printed on acid-free paper
Version Date: 20171218

International Standard Book Number-13: 978-1-4987-0973-6 (Hardback)

Library of Congress Cataloging-in-Publication Data

Names: Emani, Chandrakanth, editor.
Title: The biology of plant-insect interactions : a compendium for the plant biotechnologist / editor, Chandrakanth Emani, Department of Biology, Western Kentucky University--Owensboro, Owensboro, KY, USA.
Description: Boca Raton, FL : CRC Press, Taylor & Francis Group, 2018. | "A science publishers book." | Includes bibliographical references and index.
Identifiers: LCCN 2017050111 | ISBN 9781498709736 (hardback)
Subjects: LCSH: Insect-plant relationships. | Plants--Insect resistance. | Beneficial insects. | Insects--Effect of global warming on. | Plants--Effect of global warming on. | Plants--Disease and pest resistance.
Classification: LCC SB993 .B56 2018 | DDC 595.717/8--dc23
LC record available at https://lccn.loc.gov/2017050111

Visit the Taylor & Francis Web site at
http://www.taylorandfrancis.com

and the CRC Press Web site at
http://www.crcpress.com

Preface

Insect is derived from Latin "insectum" means a "notched or divided body" and more literally is translated as "cut into". Though the term "cut into" refers more to the notched or cut up body segments of the insect, the word can be more apt referring to the herbivory that affects plants and crop systems in the ecological realm. Since the dawn of humanity, the over million insect species that have been described have been the most significant pests that affected crops, forests and fields, in some cases affecting mass human migrations. Some of the worst documented human catastrophes since the dawn of civilization were of insect pest attacks as in the grasshopper plagues of ancient Egypt, the devastating aphid attacks that wiped out French vineyards, and the Colorado beetle attack that caused the Irish potato famine.

The oldest fossils of insects as Tuft University entomologists recovered were 300 million year old carboniferous flying insect fossil including the definitive 396 million year old "Rhynie Chert" fossil that had an insect fossil related to the modern day silverfish insect speaks of their ancient evolution that paralleled those of plants. The diversity of the insect species in multiple ecological niches of the biosphere that followed showcases the innate ability to adjust in any kind of environment. This is the reason why almost half the biosphere's organisms are insects living in environments as varied as blazing deserts to tropical forests to the polar ice caps. It is no exaggeration to state that insects have literally gobbled up more food than humans themselves consumed since agricultural production began 10,000 years ago. This led to humans devising strategies to control the insect pests. What began with environmental friendly practices such as the flooding of rice fields by south India's farmers to drown insects, or the Chinese method of infesting lemon orchards with ants that ate the Vanessa butterflies led to the usage of vegetable derived nicotine or pyrethrum extracts followed by chemicals such as arsenic and copper sulfate opening the doors to insecticides and pesticides. The research that followed in the world of chemical agriculture reached its pinnacle with the Nobel Prize winning discovery of an effective insecticide, namely, the dichloro-diphenyl-trichloro ethane (DDT) by the Swiss chemist Herman Mueller.

The usage of DDT during second world war against the malaria mosquitoes extended to agricultural fields as it was seen as a holistic extermination tool against a vast array of plant insect pests. The initial euphoria in the form of increased agricultural yields was replaced by the nightmarish environmental contamination effects as DDT was found to be harmful to mammals, fishes, birds and humans. Entomologists since that time have turned their attention to research on the actual plant-insect interactions to design more holistic strategies that ultimately led to integrated pest management. The modern era of agriculture that saw the revolutions started by plant breeders later leading to the plant genetic engineering approaches primed the entomological researchers to explore the facets of plant-insect interactions in completely newer paradigms that spanned biochemistry, genetics and molecular biology.

The present volume seeks to review the biology of plant-insect interactions as a compendium to aid the plant biotechnologist bringing together the latest advances in the field in a comprehensive fashion covering the biochemical, genetic, molecular aspects with specific case studies in model crops. The book also touches on the more recent approach of exploring the plant-insect interactions in the climate change paradigm that offers a fresh approach to the time-tested strategy of integrated pest management.

It is hoped that the book will serve the needs of not only the plant biotechnologist, but could also serve as a ready reference to plant physiologists, biochemists, entomologists in both teaching and research endeavors.

Dr. Chandrakanth Emani

Contents

CHAPTER 1

Plant Protease Inhibitors and their Interactions with Insect Gut Proteinases

S.MD. Akbar,[1,2] *Jagdish Jaba,*[1] *Visweshwar Regode,*[1] *G. Siva Kumar*[1] and *H.C. Sharma*[1,3,*]

INTRODUCTION

Enzymes hydrolysing peptide bonds have some overlapping terms, these include, proteases, proteinases and peptidases (Barrett et al. 1998). The Nomenclature Committee of the International Union of Biochemistry and Molecular Biology (NC-IUBMB 1992) recommended peptidase as the general term for enzymes hydrolyzing peptide bonds, which is further divided into exopeptidases, which catalyse the cleavage of one or a few amino acids from N-/C-terminus, and endopeptidases, which cleave the internal peptide bonds of polypeptides. The term "protease" includes both exopeptidases and endopeptidases while "proteinase" designates only endopeptidases (Barrett et al. 1998). Proteolytic enzymes are extensively found in plants, animals and in microorganisms (Kenny 1999) with a major role involved in every aspect of their physiology and development. Proteases are highly specific to their substrate, and the specificity depends on the localization of the substrate and the proteolytic enzyme, and structural and chemical properties at the active site of the enzyme.

[1] Entomology, International Crops Research Institute for the Semi-Arid Tropics (ICRISAT), Patancheru 502324, Telangana, India.
Email: akbar.pg@gmail.com; jaba.jagdish@gmail.com; vissu4@gmail.com ; siva77brinda@gmail.com
[2] Biochemistry, Agricultural Research Station, University of Agricultural Sciences, Dharwad Farm, Dharwad 580007, Karnataka, India.
[3] Dr. Y.S. Parmar University of Horticulture & Forestry, Nauni 173230, Solan, Himachal Pradesh, India.
* Corresponding author: hcsharma@yspuniversity.ac.in

Their mode of action varies among all families and groups of proteases. Some of them work individually, some work in cascades in cooperation with other proteases and some form complexes constituting an active proteolytic machine. In plants, various roles of proteolytic enzymes involves: removal of misfolded, modified, and/or mistargeted proteins; supply of amino acids during translation; maturation of zymogens and peptide hormones by partial cleavages; control of metabolism and homeostasis by altering the levels of key enzymes and regulatory proteins; and the cleavage of targeted signals from proteins prior to their final integration into organelles (Vierstra 1996). In insects, proteolysis allows digestion of wide range of food diet mediated by concerted action of several proteases and several of them such as trypsin, chymotrypsin, aminopeptidase, etc., have been characterized from a vast variety of insect pests till now (Anwar and Saleemuddin 2002; Sanatan et al. 2013; Akbar et al. 2017). The insect attack on plants triggers the production of a series of secondary metabolites; definsins, thionines, lectins, and protease inhibitors which altogether constitute the defensive armoury of plants (Buchmanan et al. 2002). Plant protease inhibitors are proteinacious in nature and inhibit insect gut proteases by binding tightly to the active site, forming an essentially irreversible complex. The inability to utilize ingested protein and to recycle digestive enzymes results in critical amino acid deficiency, which affects the growth, development and survival of the herbivore (Chougule et al. 2008). In this chapter, we aim to summarize the interactions between insect midgut proteases and the plant protease inhibitors induced as a result of insect attack.

Plant Protease Inhibitors

Plant proteinase inhibitors (PIs) play major role as potent defensive proteins against insect pests and pathogens. Diverse endogenous functions for these proteins has been proposed ranging from regulators of endogenous proteinases to storage proteins, but evidences for many of these roles is partial, or confined to limited examples. Plant PIs are small proteins generally present in high concentration in storage tissues, contributing up to 10% of the total protein content, they are also detectable in leaves in response to the attack of insects and pathogenic microorganisms (Ryan 1990). Many PIs act as defensive compounds against major insect pests either by direct assay or by expression in transgenic crop plants. The genes responsible for the production of PIs have been deployed in plants for the building of transgenic crop plants as a part of implications in integrated pest management programmes. Plant PIs are one of the important candidates with highly proven inhibitory activity against insect pests and also known to improve the nutritional quality of food (Sharma et al. 2015). The role of PIs for plant protection was investigated as early as 1947, when Mickel and Standish (1947) ascertained that the larvae of few insects were unable to complete its biology normally on soybean products. Subsequently, the trypsin inhibitors present in soybean were observed to be toxic to the larvae of flour beetle *Tribolium confusum* (Lipker et al. 1954). The Protease inhibitor is the largest class of proteins that have undergone extensive investigations and consequently, their structure, properties, function and metabolism have been well documented. *In vitro* feeding trials using artificial diets containing the inhibitors have confirmed the protective role for protease inhibitors against several crop pests.

The detrimental effects of plant PIs to insect pests are accomplished by blocking insect midgut proteinases resulting in impaired protein digestion, which inhibits or at least delays (in the case of weak inhibitors) the release of peptides and amino acids from dietary protein. This impaired protein digestion therefore affects the insects' survival leading to lower growth rate and extended developmental period. The presence of inhibitor avoids the availability of nutrients in insects particularly sulphur containing amino acids, and thereby resulting in weak and stunted growth and in some cases it ultimately results in death (Gatehouse et al. 1992). The majority of proteinase inhibitors studied in plant kingdom originates from three main families namely Leguminosae, Solanaceae and Gramineae (Richardson 1991). These protease inhibitor genes have practical advantages over genes encoding for complex pathways, i.e., by transferring a single defensive gene from one plant species to another and expressing them from their own inducible or constitutive promoters thereby imparting resistance against insect pests (Boulter 1993). This was first demonstrated by Hilder et al. (1987) by transferring trypsin inhibitor gene from *Vigna unguiculata* to tobacco, which conferred resistance to wide range of insect pests including lepidopterans, such as *Heliothis* and *Spodoptera*, coleopterans such as *Diabrotica*, *Anthonomnous* and orthoptera such as Locusts.

Classification of Protease Inhibitors

PIs have been sorted into families, subfamilies and class based on sequence and reactive active site of the inhibitory domains. Based on sequence homologies of inhibitor domains, PIs have been classified into 48 families (Rawlings et al. 2004b). These inhibitor families have been found specific for each of the classes of proteolytic enzymes.

Serpin Family (Serine Protease Inhibitors)

The serpin family is the largest and the most widespread super family of PIs. Serpin-like genes have been identified in nearly all types of organisms, including viruses, bacteria, plants and animals (Law et al. 2006; Gettins 2002). Prokaryotes generally have a single serpin gene (Irving et al. 2002a). Inhibitors of serine proteinases have been described in many plant species, and are universal throughout the plant kingdom, with trypsin inhibitors being the most common type. Plants serpins are irreversible 'suicide' inhibitors and possess the molecular mass in the range of 39–43 kDa. The cleavage of an appropriate peptide bond in the reactive center loop of the inhibitor triggers a rapid conformational change so that catalysis does not proceed beyond the formation of an acyl-enzyme complex (Huntington et al. 2000). The majority of serpins inhibit serine proteases, but serpins that inhibit caspases (Ray et al. 1992) and papain-like cysteine proteases (Irving et al. 2002b) have also been reported. Plant serpins exhibit differing and mixed specificities towards proteases (Al-Khunaizi et al. 2002). Barley (*Hordeum vulgare*) serpin is a potent inhibitor of trypsin and chymotrypsin at overlapping reactive sites (Dahl et al. 1996a). Wheat (*Triticum aestivum*) serpins inhibit chymotrypsin and cathepsin G and have glutamic acid, lysine or arginine at P1 site (Roberts et al. 2003). Two oats (*Avena sativa*) serpins show specificity for chymotrypsin and/or elastase, and another one has specificity for trypsin and chymotrypsin at overlapping loop sites (Irving et al. 2002b,c). Squash serpin Cmps-1 also inhibits elastase at two overlapping sites (Ligoxygakis et al. 2003).

Plants do not use serine proteinases in processes involving large-scale protein digestion, and hence the presence of significant quantities of serpins with specificity towards these enzymes in plants cannot be used for the purpose of regulating endogenous proteinase activity (Reeck et al. 1997). Part of this bias can be accounted for by the fact that, mammalian trypsin is readily available and is the easiest of all the proteinases to assay using synthetic substrates, and hence is used in screening procedures. Because of these reasons the members of the serine class of proteinases have been the subject of intense research than any other class of PIs. All serine inhibitor families from plants are competitive inhibitors and all of them inhibit proteinases with a similar standard mechanism (Laskowski et al. 1980).

Serine proteinase inhibitors have anti-nutritional effect against several lepidopteran insect species (Applebaum 1985). Serine proteinases present in the midgut of insects, particularly those of Lepidoptera, were found to be inhibited by these serpins (Houseman et al. 1989). Yoo et al. (2000) have reported that feeding of purified serpin to aphids had no impact on insect survival. These data suggest a more complex role for plant serpins other than insect pests defense mechanisms (Wieczorek et al. 1985; Yoo et al. 2000). Purified Bowman-Birk trypsin inhibitor (Brovosky 1986) at 5% of the diet inhibited growth of these larvae but SBTI (Kuntiz 1945), another inhibitor of bovine trypsin, was less effective when fed at the same levels. Broadway and Duffey (1986a) compared the effects of purified SBTI and potato inhibitor II (an inhibitor of both trypsin and chymotrypsin) on the growth and digestive physiology of larvae of *Heliothis zea* and *Spodoptera exigua* and demonstrated that growth of larvae was inhibited at 10% of the proteins in their diet. Trypsin inhibitors at 10% of the diet were toxic to larvae of the *Callosobruchus maculatus* (Gatehouse and Boutler 1983) and *Manduca sexta* (Shulke and Murdock 1983). Apart from their role in defense response to insect attack, serine protease inhibitors have also shown specificity to serine proteinases of broad range of organisms. Three pure trypsin inhibitors, SBTI, LBI and an egg white inhibitor (EWI) inhibited trypsins and chymotrypsins from 12 animal species with the wide range of variability (Sharma 2015). The buckwheat (*Fagopyrum sculentum*) trypsin/chymotrypsin inhibitor interferes with spore germination and mycelium growth of the tobacco brown-spot fungus *Alternaria alternata* (Dunaevskii et al. 1997). PIs from pearl millet inhibit growth of many pathogenic fungi including *Trichoderma reesei* (Joshi et al. 1998). Inhibitors that specifically inhibit proteolytic enzymes from microorganisms and not digestive proteases of animals are common in the plant, especially legume seeds. The inhibitors of the serine class of the enzymes secreted by *Bacillus subtilis* (subtilisins, or SIs) are found in seeds or vegetative tissues of many legume, cereal, and tuberous crops.

Structural analysis of serine protease inhibitors would greatly help in enzyme engineering of the native PIs to a potent form against the target pest species than the native PIs. X-ray crystallography structure proposed for the winged bean, *Psophocarpus tetragonolobus* Kunitz-type double headed alpha-chymotrypsin inhibitor showed 12 anti-parallel beta strands joined in a form of beta-trefoil with two reactive site regions (Asn 38-Leu 43 and Gln 63-Phe 68) at the external loops (Mukhopadhyay 2000). Indian finger millet (*Eleusine coracana*) bifunctional inhibitor of alpha-amylase/trypsin with 122 amino acids has shown five disulfide bridges and a trypsin binding loop (Gourinath et al. 2000).

Bowman-Birk Inhibitors (BBIs) Family

On the basis of sequence homology, this forms another family of serine PIs. The family is named after D.E. Bowman and Y. Birk, who were the first to identify and characterize a member of this family from soybean

(*Glycine max*) (Bowman 1946; Birk et al. 1963). The soybean inhibitor is now the most-well-studied member of this family and is often referred as the classic BBIs. The inhibitors have been found in legumes (Laing and McManus 2002; Tanaka et al. 1997) and in the grass family Poaceae (Odani et al. 1986). The inhibitors of this family are generally found in seeds, but are also wound-inducible in leaves (Eckelkamp 1993). They have been classified on the basis of their structural features and inhibitor characteristics. The BBIs from dicotyledonous plants consist of a single polypeptide chain with the molecular mass of 8 kDa. These are double-headed with two homologous domains each bearing a separate reactive site for the cognate proteases. These inhibitors interact independently, but simultaneously with two proteases which may be same or different (Raj et al. 2002; Birk 1985). The first reactive site in these inhibitors is usually specific for trypsin, chymotrypsin and elastase (Qi et al. 2005). The active site configuration in these inhibitors is stabilized by the presence of seven conserved disulfide bonds (Chen et al. 1992; Lin et al. 1993).

The BBIs are cysteine-rich proteins with inhibitory activity against proteases and are widely distributed in monocot and dicot species (Lin et al. 2006). BBIs from monocotyledonous plants are of two types. One group consists of a single polypeptide chain with a molecular mass of about 8 kDa having a single reactive site. Another group has a molecular mass of 16 kDa with two reactive sites (Tashiro et al. 1987, 1990; Prakash et al. 1996). It has been suggested that larger inhibitors have arisen from smaller ones by gene duplication (Odani et al. 1986). In the case of double-headed BBIs, it has been found that the relative affinity of binding of proteases is altered when one site is already occupied. Peanut (*Arachis hypogoea*) inhibitor has been found to exhibit no activity against chymotrypsin when preoccupied with trypsin and vice versa (Tur et al. 1972). In the same way, the activity of soybean BBIs decreases 100-fold when trypsin is bound at the other site (Gladysheva et al. 1999). The BBIs family of protease inhibitors contains a unique disulfide-linked nine-residue loop that adopts a characteristic canonical conformation (Bode and Hubr 1992). The loop is called protease-binding loop and binds the protease in a substrate-like manner (Lee and Lin 1995).

The trypsin subclass of serine protease inhibitors from legume seeds exhibit insecticidal effects against several crop pests belonging to the orders of Lepidoptera, Coleoptera and Orthoptera. Many of these inhibitors are products of multigene families with varying specificities towards different proteases. The cowpea trypsin inhibitor constitutes a larger gene family of four major iso-inhibitors. Three of the iso-inhibitors are specific for trypsin at each active site and the fourth is a trypsin-chymotrypsin bifunctional inhibitor (Sharma 2015).

Kunitz Family

The Kunitz inhibitors are the second major family of inhibitors which are widely distributed and often very abundant in seeds of leguminous plants and also occurs in other groups of plants including cereal seeds. On the basis of sequence homologies Kunitz-type inhibitors form a separate family. The members of this family are mostly active against serine proteases and may also inhibit other proteases (Laing and McManus 2002; Ritonja et al. 1990). The typical legume proteins are trypsin inhibitors of M_r about 21,000 Da with four cysteine residues that form two intra chain disulphide bonds and possess one reactive site. However, in the members of the legume sub family Mimosoideae, a proteolytic cleavage occurs between the third and fourth cysteine residues resulting in a heterodimeric protein comprising chains of M_r about 5,000 and 16,000 Da linked by a single disulphide bond. The inhibitors in this family are widespread in plants and have been described in legumes, cereals and in solanaceous species (Ishikawa et al. 1994; Laskowski and Kato 1980). A 20.5 kDa Kunitz-type trypsin inhibitor with antifungal activity has been reported from the roots of punce ginseng, *Pseudostellaria heterophylla* (Wang and Ng 2006). Kunitz-type PIs are also produced under stress, as has been found in potato tubers, *Solanum tuberosum* (Park et al. 2005; Ledoigt et al. 2006; Plunkett et al. 1982).

The Kunitz type trypsin inhibitor inhibits trypsin through interaction with a single site on the inhibitor and that is encoded by the KTi3 gene. Specificity of trypsin inhibitor is determined by the two amino acids residues, arginine and isoleucine, at the active site of the KTi protein; these amino acids are considered essential for inhibitor function, although arginine and serine are the active site residues in other inhibitors (Sharma 2015). However, not all the kunitz related proteins of legume seeds are proteinase inhibitors. The winged bean albumin-1, storage protein from *Psophocarpus tetragonolobus* L., accounts for about 15% of the total seed protein. It comprises 175 amino acid residues with a Mr of 19,333 and contains the single disulphide bond. It shows 38% and 28% sequence similarity with Kunitz inhibitors from soyabean and winged bean, respectively, but has no inhibitory activity (Kortt et al. 1989). The members of this family are mostly active against serine proteases and have been shown to inhibit trypsin, chymotrypsin and subtilisin (Laing and McManus 2002; Park et al. 2005), but they also inhibit other proteases including the aspartic protease cathepsin D and the cysteine proteinase papain. These inhibitors are canonical and form a tight complex with the target protease, which dissociates very slowly (Ritonja et al. 1990). It has been found that potato tubers treated with elicitors, jasmonic, salicylic or arachidonic acids are able to excrete potatin and three chymotrypsin inhibitors (Ledoigt et al. 2006). Wounding and water stress promotes the secretion of two kinds of Kunitz-type PIs by potato tubers. These inhibitors are closely associated with other secreted polypeptides and would protect them against degradation by extracellular chymotrypsin-like protease. The secreted inhibitors could therefore interact with plant defense system (Valueva et al. 2001; Ledoigt et al. 2006).

Squash Inhibitors

Squash-family inhibitors have been described only in plants from many cucurbit families and form yet another family active against serine proteases (Lee and Lin 1995; Felizmenio et al. 2001). Seven serine PIs belonging to this family have been isolated and characterized from the seeds of wild cucumber, *Cyclanthera pedata* (Kuroda et al. 2001). Recently two different but inter-convertible (cis-trans isomers) inhibitors have been isolated and characterized from seeds of wax gourd, *Benincasa hispida* (Thumb) cogn. (Atiwetin et al. 2006). The members of this family consist of a small single peptide chain containing 28 to 30 amino acids with molecular mass of 3.0–3.5 kDa (Heitz et al. 2001). These inhibitors have three disulfide bridges and fold in a novel knottin structure (Hara et al. 1989). The small size of these inhibitors combined with potential activity against important biological molecules such as Hageman factor, human leucocyte elastase and cathepsin G (McWherter et al. 1989) has made them particularly attractive for studying proteinase and inhibitor interactions. Chemical synthesis of these inhibitors has created powerful tools for investigating their structural and functional relationship (Rolka et al. 1992). The structures of squash inhibitors and inhibitor-proteinase complexes have been determined by X-ray crystallography and NMR spectroscopy (Thaimattam et al. 2002).

Cereal Trypsin/Amylase Inhibitors

The members of this family have serine proteinase inhibitory activity and/or amylase inhibitory activity (Gourinath et al. 2000). A large number of inhibitors in this family have only amylase inhibitory activity; however, inhibitors from barley (*H. vulgare*), rye (*Secale cereale*) and tall fescue (*Festuca arundinacea*) are active against trypsin (Odani et al. 1983). Maize (*Zea mays*) and ragi (*Elusine coracana*) inhibitors show dual activities and can inhibit serine proteinases as well as amylase (Mahoney et al. 1984; Shivraj and Pattabiraman 1981). The cereal trypsin/amylase inhibitors consist of a single polypeptide chain containing five disulfide bonds with a molecular mass of about 13 kDa (Christeller and Liang 2005). The structure of the ragi inhibitor solved by NMR spectroscopy and that of its complex with yellow mealworm, *Tenebrio molitor* amylase by x-ray crystallography has shown that the proteinase-binding loop adopts a canonical conformation (Strobl et al. 1998).

Mustard (Sinapis) Trypsin Inhibitor (MSI)

These are small single polypeptide chain inhibitors with the molecular mass of about 7 kDa, found in the family Cruciferae and form yet another family of serine PIs (Laing and McManus 2002; Menengatti et al. 1992). These inhibitors have been isolated and characterized from a number of species including white mustard, *Sinapis alba,* and tape, *Brassica napus* (Ascenzi et al. 1999; Volpicella et al. 2000). These inhibitors are expressed in seeds during their development and are also wound-inducible (Ceci et al. 1995; De Leo et al. 2001). The inhibitors form a tight binding complex with trypsin and apparently follow the standard mechanism of inhibition as that of serpins (Ceciliani et al. 1994).

Potato Type I Protease Inhibitors (PI 1)

The inhibitors of this family are widespread in plants and have been described in many species including potato tubers (Ryan and Balls 1962), tomato fruit (Margossian et al. 1988; Wingate et al. 1989), squash phloem exudates (Murray and Christeller 1995) and in tomato leaves in response to wounding (Lee et al. 1986). These inhibitors have the molecular mass of 8 kDa and are generally monomeric. While the inhibitors from cucurbit and potato tubers contain a single disulphide bond, the inhibitors in this family lack any disulphide bridges (Cai et al. 1995). The inhibitory mechanism in this family is considered to fit the standard model.

Potato Type II Protease Inhibitors (PI 2)

The members of this group have been reported only from the members of Solanaceae family. Initially characterized from potato tubers (Christeller and Liang 2005) and these inhibitors have been found in leaves, flowers, fruit and phloem of other solanaceaous species (Iwasaki et al. 1971; Pearce et al. 1993). Among the PIs, the wound-inducible inhibitors from potato and tomato represent a unique group with insecticidal properties due to several interesting features of these proteins and their encoding genes. An analysis of these inhibitors and genes has shown that they are composed of multiple repeat units varying between one and eight (Antcheva et al. 2001; Miller et al. 2000; Choi et al. 2000). They comprise a non-homologous gene family in which members have been identified mainly from the solanaceous plants. Among them, potato inhibitor I and II, tomato protease inhibitor I and II have been well characterized. The unique and most striking feature of their encoding genes are the presence of introns, two each in inhibitor I genes and one in the gene encoding potato inhibitor II. In fact, they are the only protease inhibitor genes reported so far to contain introns. In potato alone, a mixture of one or more inhibitors of protease inhibitor I and at least three forms of inhibitor II have identified. In addition, homologs of the inhibitor have been found in some non-solanaceous plants like alfalfa, broad bean, clover, cowpea, cucumber, french bean, grape, squash, strawberry, barley and buckwheat.

In leaves of tomato and potato, they are expressed constitutively at low levels during plant growth and development. In response to wounding by insects or other mechanical damage, their concentration increases dramatically even in the unwounded leaves of the same plant and within a few hours of injury, their levels often exceed 10% of total soluble proteins. In potato tubers, they accumulate throughout the course of tuber development and represent a substantial fraction of the soluble protein. Thus, unlike other plant protease inhibitor gene, these genes are regulated environmentally as well as developmentally and their expression is believed to be under a complex control involving several *cis* and *trans*-acting factors making them excellent models for study of plant gene regulation. A low molecular-mass inhibitor of this family has been found to be constitutively present in Jasmme tobacco (*Nicotiana alata*) flowers (Atkinson et al. 1993). Six small wound-inducible proteinase inhibitors of this family have been reported from tobacco leaves (Pearce et al. 1993). Inhibitors of this family have been reported to inhibit chymotrypsin, trypsin, elastase, oryzin, Pronase E and subtilisin (Antcheva et al. 2001).

Cysteine Protease Inhibitors (The Cystatin Superfamily)

The members of these families inhibit the activity of cysteine proteases and are called cysteine PIs or cystatins. The cystatin super family (CYS) is composed of several families and includes proteins that are related in structure and function to cysteine proteinase inhibitors. They were first described in egg white and referred to as 'chicken eggwhite cystatin' (Colella et al. 1989). They are widely distributed in plants, animals and microorganisms (Oliveira et al. 2003). Most cysteine proteinase inhibitors have been found in animals, but several have been isolated from plant species as well including pineapple, potato, corn, rice, cowpea, mungbean, tomato, wheat, barley, rye and millet. Cysteine proteinases are not secreted as intestinal digestive enzymes in higher animals, but are found in the midguts of several families of Hemiptera and Coleoptera where they appear to play important roles in the digestion of food proteins. In a study of the proteinases from the midguts of several members of the order coleoptera, 10 of 11 beetle species representing 11 different families had gut proteinases that were inhibited by p-chloromercuribenzene sulphonic acid (PCMBS), a potent sulphydryl reagent, indicates that the proteinases are in the pH range of the insect gut that usually possess cysteine proteinases. Expression of the PI genes of these inhibitors are usually limited to specific organs or to particular phases during plant growth, such as germination (Botella et al. 1996), early leaf senescence (Huang et al. 2001) and drought (Van der-Vyver et al. 2003; Pernas et al. 2000). Wounding or treatment with methyl jasmonate evokes a similar pattern of gene expression. Further, the cytosolic localization of these inhibitors also suggests that they are involved in plant defense against insects (Zhao et al. 1996).

The rice cysteine proteinase inhibitors are the most studied of all the cysteine PIs which is proteinaceous in nature and highly heat stable (Abe et al. 1987). Phytostatins from various plants inhibit the activity of gut cysteine proteinases involved in protein digestion in the gut of various members of the Coleoptera (beetles) attacking these plants, and thus play a role in the exogenous defense system of these plants (Oliveira et al. 2003). *Oryza cystatin* is found to prevent the growth of rice weevil, *Sitophilus oryzae*, by inhibiting the cysteine proteases in the gut of this organism (Hosoyama et al. 1994). The rice cystatins have been reported to confer resistance against potyviruses in transgenic tobacco and sweet-potato plants (Campos et al. 1999). Two extracellular cysteine protease inhibitors (ECIP-1 and ECIP-2) isolated from species of the unicellular green alga *Chlorella* seem to have a role in protecting the cells from attacks by viruses and insects (Ishihara et al. 1999, 2000). Cystatins have also been characterized from potato, ragweed, cowpea, papaya and avocado. Cysteine PIs from pearl millet (*Pennisetum glaucum*) inhibits the growth of many pathogenic fungi, including *Trichoderma reesei* (Joshi et al. 1998). These advantages make protease inhibitors an ideal choice to be used in developing transgenic crops resistant to insect pests and pathogen. Transgenic rice expressing maize cystatin has been shown to exhibit enhanced resistance against insect predation (Irie 1996).

Zeins and maize proteinases are inhibited by maize cystatins, suggesting a role for these inhibitors in the endogenous defense mechanism (Steller 1995; Hoorn and Jones 2004). Phytostatins are involved in the control of endogenous cysteine proteinases during maturation and germination of seeds (Abe and Arai 1991) and play a role in the apoptosis required in plant development and senescence (Solomon et al. 1999). Oryzacystatins have been shown to inhibit the cysteine proteinases that are produced during seed germination (Watanabe et al. 1991).

The over-expression of cystatin in soybean cell suspensions blocked programmed cell death (PCD) (Solomon et al. 1999). The over-expression of a cystatin that inhibits papain activity in *Arabidopsis* cell cultures blocked cell death in response to avirulent bacteria and nitric oxide (Belenghi et al. 2003). The over-expression of this inhibitor in tobacco plants blocked the hypersensitive response induced by avirulent bacteria (Hoorn and Jones 2004; Belenghi et al. 2003). These cysteine protease inhibitors are grouped into four families based on sequence relationships, molecular mass and disulfide-bond numbers and arrangements (Barrett 1987).

Family-1 Cystatins (Stefin Family)

The members of this group have a molecular mass of about 11 kDa. They are generally present in the cytosol and are devoid of any carbohydrate groups and disulfide bonds (Stato et al. 1990; Machleidt et al. 1983).

Family-2 Cystatins (Cystatin Family)

These inhibitors consist of proteins having 120–126 amino acids with molecular mass of 13.4–14.4 kDa. These inhibitors contain two disulphide bonds but are devoid of any carbohydrate groups (Grzonka et al. 2001). They also contain a signal sequence and are known to be secreted (Abrahanson et al. 1987). All the family-2 cystatin inhibitors contain a conserved tripeptide sequence, Phe-Ala-Val near the C-terminus, and a conserved dipeptide, Phe-Tyr, near the N-terminus. These conserved sequences are important in binding to the target proteases (Machleidt et al. 1983; Turk et al. 1997).

Family-3 Cystatins (Kininogen Family)

These inhibitors are glycoproteins and are of three different types, High Molecular Weight kininogens (HMW) with a molecular mass of 120 kDa, Low Molecular Weight kininogens (LMW) with molecular mass ranging between 60 and 80 kDa, and a third type, T kininogens with a molecular mass of 68 kDa. These proteins contain tandem domains that result from gene duplication of the family-2 cystatins. These proteins are also secreted and play key roles in blood coagulation (Otto and Schirmeister 1997; Salvesen et al. 1986). Family 1 and 3 cystatins contain a conserved pentapeptide sequence, Gln-Val-Val-Ala-Gly, and the family-2 members have the homologous peptide, Gln-X-Val-Y-Gly, in which X and Y represent any amino acid (Habib and Fazilil 2007).

Family-4 Cystatins (Phytocystatins)

This family includes nearly all the cysteine PIs described in plants. They have been identified in rice (Abe et al. 1987a,b), maize (Abe et al. 1992), soybean (Hines et al. 1991; Botella et al. 1996), apple (*Malus*) fruit (Ryan et al. 1998), carnation (*Dianthus caryophyllus*) leaves (Kim et al. 1999) and several other monocotyledonous and dicotyledonous plants (Brown and Dziegielewska 1997; Pernas et al. 1998; Sakuta et al. 2001). Celostatin, a cysteine PI from crested cock's comb (*Celosia cristata*) has recently been cloned and characterized (Gholizadeh et al. 2005). Phytocystatins have sequence similarity to stefins and cystatins, but do not contain free cysteine residues (Fernandes et al. 1993; Zhao et al. 1996). However, the unique feature of this superfamily is a highly conserved region of the G58 residue, the glu-val-val-ala-gly (QVVAG) motif and a pro-trp (PW) motif. The studies on the papain inhibitory activity of Oryzacystatin and its various truncated forms have identified the conserved QVVAG motif as a primary region of interaction between the inhibitor and its cognate enzyme. The PW motif is believed to act as a cofactor (Arai et al. 1991; Abe et al. 1988). Phytocystatins, based on protein structure, have been divided into two groups, one group consists of single-domain proteins and includes most these inhibitors (Abe et al. 1987a,b; Pernas et al. 1998), another group contains multiple-domains and includes the cysteine PIs isolated from potato tubers and tomato leaves (Walsh and Strictland 1993; Bolter 1993). Plant cysteine PIs are encoded by gene families (Fernandes et al. 1993) and show different expression patterns during development and defense response to biotic and abiotic environmental stress (Felton and Korth 2000). The expression is usually limited to specific organs or to specific phases during development, such as germination (Botella et al. 1996), early leaf senescence (Huang et al. 2001) cold and salt stress (Van der-Vyver et al. 2003; Pernas et al. 2000).

Aspartyl and Metallocarboxypeptidase Inhibitors

Aspartyl PIs have been isolated from sunflower, barley and cardoon (*Cynara cardunculus*) flowers are named as cardosin A (Park et al. 2000; Kervinen et al. 1999; Lawrence and Koundal 2002; Mares et al. 1989; Wolfson and Murdock 1987). In species of six families of the order Hemiptera, aspartic proteinases (cathepsin D-like) were found along with cysteine proteinases (Houseman 1983). The cathepsin D inhibitor, an aspartyl PI described in potato tubers shares considerable amino acid sequence homology with soybean trypsin inhibitor. It is a 27 kDa protein and inhibits serine proteases trypsin and chymotrypsin in addition to the aspartyl protease cathepsin D, but does not inhibit pepsin, cathepsin E and rennin, which are all aspartyl proteases (Lawrence and Koundal 2002). The inhibitor also accumulates in potato leaf tissues along with serine proteinase inhibitor I and II proteins in response to wounding. Thus, the inhibitors accumulated in the wounded leaf tissues of potato have the capacity to inhibit all the five major digestive enzymes, i.e., trypsin, chymotrypsin, elastase, carboxypeptidase A and carboxypeptidase B of higher animals and many insects (Hollander 1985). Pepstatin, a powerful and strong inhibitor of aspartyl proteases has been shown to inhibit proteolysis of the midgut enzymes of Colorado potato beetle (*Leptinotarsa decemlineata*) (Wolfson and Murdock 1987).

Plants contain two families of metalloproteinase inhibitors, the metallocarboxypeptidase inhibitor family in potato and tomato plants (Graham and Ryan 1997; Rancour and Ryan 1968) and a cathepsin D inhibitor family in potatoes (Keilova and Tomasek 1976). The inhibitors that bind to metallocarboxypeptidases have been identified in solanaceous plants, in the medicinal leech (*Hirudo medicinalis*), in the intestinal parasite roundworm *Ascaris sum*, in the blood tick *Rhipicephalus bursa* and in rat and human tissues (Arolas et al. 2005; Homandberg et al. 1989; Reverter et al. 1998; Normant et al. 1995; Liu et al. 2000). These inhibitors are small peptide inhibitors consisting of 38–39 amino acid residues and have the molecular mass of about 4.2 kDa (Hass et al. 1975; Hass and Hermodson 1981). These inhibitors are polypeptides (4 kDa) inhibit strongly but competitively to a broad spectrum of carboxypeptidases from both animals and microorganisms, but do not inhibit serine carboxypeptidases from yeast and plants (Havkioja and Neuvonen 1985). A metallocarboxypeptidase inhibitor is found to accumulate in potato tuber tissues during development, along with the potato inhibitor I and II families of serine PIs. The inhibitor also accumulates in potato leaf tissues, along with the inhibitors of other families, as a response to wounding (Ryan 1990).

Insect Resistant Transgenic Plants Expressing PIs

Since the economically important orders of insect pests namely Lepidoptera, Diptera and Coleoptera use serine and cysteine proteinases in their digestive system to degrade proteins in the ingested food and efforts have generally been directed at genes encoding PIs active against these mechanistic classes of proteases for developing transgenic plants. The PI genes have been particularly utilized in developing transgenic plants resistant to insect pests and/or pathogen by transferring a single defensive gene that can be expressed from the wound-inducible or constitutive promoters of the host (Boulter 1993). Several transgenic plants expressing PIs have been produced in the last 15 years and tested for enhanced defensive capabilities with particular efforts directed against insect pests (Valueva et al. 2001).

The PI gene coding for cowpea trypsin inhibitor (CpTi) was the first to be successfully transferred and produced transgenic tobacco with significant resistance against tobacco hornworm, *Manduca sexta* (Hilder et al. 1987). The efficiency of transgenic tobacco plants expressing CpTi was also tested against armyworm, *Spodoptera litura,* in feeding trials under laboratory conditions. Reduction of 50% biomass was observed in the larvae fed on transgenic leaves expressing 3–5 μg of CpTi/g of fresh leaves (Sane et al. 1997). Potato PI-II gene from potato was introduced into several japonica rice varieties to produce transgenic rice plants shown to be insect resistant in greenhouse trials. Wound-inducible PI-II promoter with the first intron of rice actin I gene could give high-level expression of PI-II gene in transgenic rice plants were resistant to pink stem borer, *Sesamia inferens* (Duan et al. 1996).

Bean α-amylase inhibitor I in transgenic peas, *Pisum sativum,* provided complete protection from pea weevil, *Bruchus pisorum,* under field conditions (Roger et al. 2000). When both soybean BBI and Kunitz inhibitors were introduced and expressed in sugarcane, *Saccharum officinarum,* the growth of neonate larvae of sugarcane borer, *Diatraea saccharalis* feeding the leaf tissues was significantly retarded as compared to larvae feeding on leaf tissues from untransformed plants (Falco and Silva 2003). The transgenic wheat, *T. aestivum,* carrying barley trypsin inhibitor gene (BTI) showed a significant reduction of infestation with Angoumois grain moth, *Sitotroga cerealella.* However, only early-instar larvae were inhibited in transgenic seeds and expression of BTI protein in transgenic leaves did not have a significant protective effect against leaf-feeding insects (Altpeter et al. 1999).

The PIs also exhibited a very broad spectrum of activity against pathogenic nematodes. CpTi inhibited the growth of nematodes, *Globodera tabacum, G. pallida* and *Meloidogyne incognita* (Williamson and Hussey 1996). Transgenic potato expressing two cystatin genes developed resistance to a nematode, coleopteran insects (Cowgill et al. 2002) and transgenic rape plants expressing rice cystatin 1 were resistant to aphid (Rahbe et al. 2003). Recently, protease inhibitors have also been used to engineer resistance against viruses in transgenic plants (Ussuf et al. 2001). The presence of anti-fungal and anti-feeding activity on a single protein explored a new possibility of raising a transgenic plant resistant to pathogens, as well as pests by transfer of a single CPI gene. Pearl millet cysteine protease inhibitor (CPI) has been found to possess anti-fungal activity in addition to its antifeedant activity against insects (Joshi et al. 1998). Expression of Oryzacystatin, the rice cysteine proteinase inhibitor, into the tobacco plant induced significant resistance against two important potyviruses, tobacco etch virus (TEV) and potato virus Y (PVY). These results suggest that plant cystatins can be used against different potyviruses and potentially also against other viruses whose replication involves cysteine proteinase activity (Campos et al. 1999).

These advantages make protease inhibitors an ideal choice to be used in developing transgenic crops resistant to insect pests. Further, the transformation of plant genomes with PI-encoding cDNA clones appears attractive not only for the control of plant pests and pathogens, but also as a means to produce PIs, useful in alternative systems and the use of plants as factories for the production of heterologous proteins. A list of transgenics expressing plant PIs are included in Table 1.

Plant PIs in Defense and Limited Success as Insecticides

Plants PIs vary in protein primary sequence and tertiary structure to act with all mechanistic protease groups. A new PI categorization system has been showed with the increasing availability of sequence data and 3-D structural information, gradually replacing the previous categorization based on protease specificity (Rawlings et al. 2004; Jongsma and Beekwilder 2011). Although many PIs are minor proteins having a single inhibitory domain, it is not uncommon for PIs to contain two or more inhibitor units. Potato multi cystatin, for example, has eight tandem cystatin domains (Walsh and Strictland 1993).

The general mode of action of PI molecules includes inhibition of protein digestive enzymes in insect guts resulting in amino acid deficiencies which lead to delayed developmental growth, increased mortality, and/or reduced fecundity (Gatehouse 2011). The adverse effects of dietary PIs on insects may be more complex actions than a simple decrease in the proteolytic activity of the digestive enzymes complex. Feedback mechanisms in response to dietary PIs were suggested to lead to the hyper production of proteases to counterbalance for the loss of activity, causing the declining trends of essential amino acids. The imposed nutritional stress would later retard insect growth and development (Schechter and Berger 1967; Broadway and Duffey 1986). In addition to the direct inhibitory effect on proteolytic enzyme complexes, plant PIs may function in other processes. For instance, PIs inhibited normal development of cereal aphids, although a relatively higher concentration of free amino acids present in the sap it could not prevent the growth impairment caused by PIs, thus indicates the indirect effects on the aphid rather than inhibiting food protein digestion (Pyati et al. 2011).

Although plant PIs are an important component in insect pest management, but attempts to use single PIs in transgenic crops has very limited success. Since the initial effort in expressing a cowpea trypsin inhibitor in tobacco, *Nicotiana tabacum* (Hilder et al. 1987), transgenic plants holding foreign PI genes of various types have been developed in a range of plants, including wheat (*T. aestivum*), rice (*Oryza sativa*), cotton (*Gossypium hirsutum*) and *Arabidopsis thaliana* (Chapman 1988; Mosolov and

Table 1. List of few transgenic plants expressing plant PIs that have been developed and tested for their effectiveness on the growth and development of insect pests.

Protease inhibitor	Protease family	Proteases inhibited	Transformed plant	Insect species used in bioassay	Effect of PI on larval growth
Arabidopsis thaliana serpin 1 [AtSerpin1]	alpha-1-peptidase inhibitor	Chymotrypsin	*Arabidopsis*	*Spodoptera littoralis*	38% biomass reduction after feeding for 4 days (Alvarez-Alfageme et al. 2011)
Barley trypsin inhibitor [BTI]	Cereal Trypsin inhibitor	Trypsin	Tobacco	*Spodoptera exigua*	29% reduction in survival (Altpeter et al. 1999)
			Wheat	*Sitotroga cerealella*	No effect on growth or mortality (Lara et al. 2000)
Bovine pancreatic trypsin inhibitor [BPTI]	Kunitz (animal)	Trypsin, chymotrypsin, plasmin, kallikreins	Tobacco	*Spodoptera exigua*	Reduced trypsin activity; induced leucine, aminopeptidase and carboxypeptidase A activities; chymotrypsin, elastase, and carboxypeptidase B proteases not affected (Lara et al. 2000)
			Sugarcane	*Scirpophaga excerptalis*	Significant reduction in weight (Christy et al. 2009)
Bovine spleen trypsin inhibitor [SI]	Kunitz (animal)	Trypsin, chymotrypsin	Tobacco	*Helicoverpa armigera*	Reduced survival and growth (Christy et al. 2009)

Table 1 contd. ...

...Table 1 contd.

Protease inhibitor	Protease family	Proteases inhibited	Transformed plant	Insect species used in bioassay	Effect of PI on larval growth
Cowpea trypsin inhibitor [CpTI]	Bowman-Birk	Trypsin	Tobacco	*Heliothis virescens*	Increased mortality (Hilder et al. 1987)
			Tobacco	*Helicoverpa zea*	Increased mortality (Hoffmann et al. 1992)
			Rice	*Chilo suppressalis-Sesamia inferens*	Growth not monitored (Xu et al. 1996)
			Potato	*Lacanobia oleracea*	45% biomass reduction (Gatehouse et al. 1997)
			Tobacco	*Spodoptera litura*	50% biomass reduction (Sane et al. 1997)
			Potato	*Lacanobia oleracea*	Decreased weight and delayed development (Bell et al. 2001)
Giant taro proteinase inhibitor [GTPI]	Kunitz (plant)	Trypsin, chymotrypsin	Tobacco	*Helicoverpa armigera*	Decreased growth, no increase in mortality (Wu et al. 1997)

Mustard trypsin inhibitor 2 [MTI-2]	Brassicaceae proteinase inhibitor	Trypsin, chymotrypsin	Tobacco, *Arabidopsis* and oilseed rape	*Spodoptera littoralis*	Increased mortality; surviving larvae up to 39% smaller after 10 days (De Leo et al. 1998)
				Mamestra brassicae, Plutella xylostella, Spodoptera littoralis	*P. xylostella:* 100% mortality on *Arabidopsis*; high mortality & delayed development on oilseed rape. *M. brassicae:* increased mortality & weight of survivers on *Arabidopsis* and tobacco, no effect on oilseed rape. *S. littoralis*: delay in development on oilseed rape (De Leo et al. 2001)
			Tobacco	*Spodoptera littoralis*	No effect on growth; reduction in fertility (De Leo and Gallerani 2002)
			Oilseed rape	*Plutella xylostella*	Reduction in survival and weight (Ferry et al. 2005)
Nicotiana alata protease inhibitor [NaPI]	Proteinase inhibitor II	Trypsin, chymotrypsin	Tobacco	*Helicoverpa punctigera*	Decreased weight; increased mortality (Heath et al. 1997)
			Tobacco and peas	*Helicoverpa armigera*	Increased mortality; delayed growth (Charity et al. 1999)
			Royal Gala' apple	*Epiphyas postvittana*	Larval and pupal weights reduced; developmental abnormalities (Maheswaran et al. 2007)
			Cotton	*Helicoverpa armigera*	A higher number of cotton bolls were recorded in plants expressing NaPI and a PotI inhibitor from potato, StPin1A (Dunse et al. 2010)

Table 1 contd. ...

...Table 1 contd.

Protease inhibitor	Protease family	Proteases inhibited	Transformed plant	Insect species used in bioassay	Effect of PI on larval growth
Potato inhibitor II [Pin II, PPI-II, Pot II, PI-II]	Proteinase inhibitor II	Trypsin, chymotrypsin, oryzin, subtilisin, elastase	Tobacco	*Manduca sexta*	Growth retarded (Johnson et al. 1989)
			Tobacco	*Chrysodeixis eriosoma, Spodoptera litura, Thysanoplusia orichalcea*	*C. eriosoma* larvae grew slower; *S. litura and T. orichalcea* growth either unaffected or enhanced (McManus et al. 1994)
			Tobacco	*Spodoptera exigua*	Growth not affected (Jongsma et al. 1995)
			Rice	*Sesamia inferens*	Decreased weight (Duan et al. 1996)
			Brassica napus	*Plutella xylostella*	Lowered growth rates however more plant tissue consumed (Winterer and Bergelson 2001)
			Tomato	*Heliothis obsoleta*	Increased mortality and decreased weight on homozygous plants expressing PI-II and potato carboxypeptidase inhibitor (PCI), opposite effect on hemizygous plants (Abdeen et al. 2005)
Solanum americanum proteinase inhibitor [SaPIN2a]	Proteinase inhibitor II	Trypsin, chymotrypsin	Tobacco	*Helicoverpa armigera, Spodoptera litura*	Reduction in larval weight and pupation rate (Luo et al. 2009)

Soybean Kunitz trypsin inhibitor [SBTI, SKTI]	Kunitz (plant)	Trypsin, chymotrypsin, kallikrein, plasmin	Poplar	*Clostera anastomosis, Lymantria dispar*	Mortality and growth not Significantly affected (Confalonie et al. 1998)
			Potato	*Lacanobia oleracea*	Survival and growth decreased by 33% and 40% respectively after 21 days (Gatehouse et al. 1999)
			Tobacco	*Spodoptera litura*	Increased mortality and Delayed development (McManus et al. 1999)
			Tobacco	*Helicoverpa armigera*	Development unaffected (Nandi et al. 1999)
			Tobacco and potato	*Spodoptera littoralis*	High mortality on tobacco and up to 50% weight reduction on potato (Marchetti et al. 2000)
			Sugarcane	*Diatraea saccharalis*	Increased mortality; retarded growth (Falco et al. 2003)
Soybean Bowman-Birk trypsin inhibitor [SBBI]	Bowman-Birk	Trypsin, chymotrypsin	Sugarcane	*Diatraea saccharalis*	Growth severely retarded (Falco et al. 2003)
			Cauliflower	*Plutella xylostella, Spodoptera litura*	Increased mortality (Duan et al. 1996)
			Tobacco	*Spodoptera litura*	Growth and survival severely retarded (Yeh et al. 1997)

Table 1 contd. ...

...Table 1 contd.

Protease inhibitor	Protease family	Proteases inhibited	Transformed plant	Insect species used in bioassay	Effect of PI on larval growth
Sweet potato trypsin inhibitor [SWTI, Sporamin]	Kunitz (plant)	Trypsin	Tobacco	*Helicoverpa armigera*	Increased mortality and delayed growth and development in larvae on plants expressing sporamin and a phytocystatin from taro, CeCPI (Senthilkumar et al. 2010)
			Brassica	*Plutella xylostella*	Survival rate and body mass was significantly lower in larvae fed plants. Expressing sporamin and chitinase (Liu et al. 2011)
Tomato inhibitor I [Tom1]	Proteinase inhibitor I	Chymotrypsin, subtilisin, trypsin	Tobacco	*Manduca sexta*	Little effect on growth (Johnson et al. 1989)
Tomato inhibitor II [TPI-II]	Proteinase inhibitor II	Chymotrypsin, trypsin, subtilisin	Tobacco	*Manduca sexta*	Growth retarded (Johnson et al. 1989)
Beta vulgaris serine proteinase inhibitor gene (BvSTI)			*Nicotiana benthamiana*	*Spodoptera frugiperda, Spodoptera exigua, Manduca sexta, Heliothis virescens* and *Agrotis ipsilon*	Reductions in larval weights in *S. frugiperda, S. exigua* and *M. sexta*. Developmental abnormalities of the pupae and emerging moths (Johnson et al. 1989)

Valueva 2008). However, no PI-transgenic plants have made commercially available, despite the detrimental effects of the PIs on insect pests. Under selection pressure, insects appear to develop resistance to many defense genes including those encoding PIs. Some even overcompensate for the nutritional hindrance by consuming more transgenic plant material and gaining more weight than with non-transgenics (De Leo et al. 1998).

The PIs undoubtedly form a significant component of a multi-mechanistic defense system used by plants. However, the remarkably flexible physiological responses displayed by insects facilitate their adaptation to the PI-based plant defense. Consequently, non-plant PIs were explored as a more effective measure to avoid the insufficiency of inhibition of plant PIs due to specific co-evolutionary interactions (Harsulkar et al. 1999). Spit et al. (2012) designed an inhibitor mixture based on the total midgut protease profile of the desert locust, *Schistocerca gregaria*. The mixture contains pacifastin-related inhibitors that are of non-plant origin with maximal *in vitro* inhibition of the trypsin- and chymotrypsin-like proteolytic activity of the *S. gregaria* midgut. Pacifastin family members have been found in all arthropods. Those that have been characterized are derived from precursor polypeptides consisting of an N-terminal signal sequence followed by a variable number of inhibitor domains (Breugelmans et al. 2009). The pacifastin-related inhibitor mixture resulted in greater suppression of insects compared with plant-derived inhibitors in feeding assays. However, the inhibitory effect gradually waned and insects recovered within a few days from the growth impediment.

A conjunct search for novel PIs apparently has resulted in little success. It has gradually become clear that the degree to which proteolytic digestion in an insect vulnerability is dependent not only on the PI's defensive mechanism but also on the insect's response. Fundamental knowledge of insect counter defense is a prerequisite for making use of PI and other anti-insect molecules for producing the next generation of insect-resistant transgenic plants.

Mechanism of PI Toxicity

The toxicity of the plant PIs depends upon the structural compatibility with the protease and the physiological conditions of gut of the target organism (e.g., pH) and also the quality and quantity of PI ingested (Broadway 1995). The mechanism of binding of the plant PIs to the insect proteases appears almost similar with all the classes of inhibitors (Laskowski et al. 1980). The inhibitor binds at the active site of the enzyme forming an enzyme-inhibitor complex with a very low dissociation constant (10^7 to 10^{14} M at neutral pH values) thereby efficiently blocking the active site. The inhibitor therefore directly mimics a normal substrate of the enzyme thereby blocking the usual enzyme mechanism of cleaving the peptide bond (Walker et al. 1998). Sometimes, a binding loop from the inhibitor projects from the surface of the molecule and contains a cleavable peptide bond for the enzyme, but the cleavage do not interfere the interaction between enzyme and the inhibitor (Terra et al. 1996; Walker et al. 1998).

Conclusion

Plant PIs are key players in the endogenous defense system as they help regulate and balance protease activities. These inhibitors are also important participants in the exogenous defense. The importance of PIs has been realized for some time now, and many transgenic plants over expressing different PIs have been produced with resistance against different insects and/or pathogenic organisms. This is, however, yet to be fully appreciated, and it can have important consequences beyond their recognized scope. These inhibitors can also interfere with the life cycle of many viruses and may help prevent many viral disorders. Although plant PIs have been isolated and characterized from many sources, and that the natural inhibitors have been made available through transgenic plants over expressing specific inhibitors with the potential for the natural inhibitors in agriculture is enormous, awaiting full-scale exploration to combat the insects' counter adaptations developed against them for sustainable crop production.

Insect Gut Proteinases

Insects generally use different types of digestive enzymes to digest a wide range of food diets, including polymeric molecules, which are secreted by midgut's epithelial cells (Terra and Ferreira 1994; Terra et al. 1996). Some carbohydrases and proteases can break down the carbohydrates and proteins into absorbable elements in midgut, respectively (Terra 1990). The proteinases are a noteworthy group of hydrolytic enzymes in insects and are included in digestive processes, proenzyme activation, freedom of physiologically dynamic peptides, supplement initiation, and aggravation forms among others (Neurath 1984). The proteinases are characterized by their mechanisms of catalysis: (1) serine proteinases; (2) cysteine proteinases; (3) aspartic proteinases; and (4) metalloproteinases (Bode and Huber 1992). For a proficient management of pest control through proteinase inhibitor transgenes, it is basic to know the sort of catalysts present in the gut of insects and pests. The two noteworthy proteinase classes in the digestive systems of phytophagous insects are the serine and cysteine proteinases (Haq et al. 2004). Murdock et al. (1987) completed a detailed investigation of the midgut proteins of different pests having a place with Coleoptera, while Srinivasan et al. (2006) have covered the midgut enzymes of different pests having a place with Lepidoptera. Serine proteases are known to command the larval gut condition and add to around 95% of the aggregate stomach related action in Lepidoptera, though the Coleopteran species have a more extensive scope of predominant gut proteinases.

Serine Proteinases

Serine proteases (SPs) in the chymotrypsin (S1) family constitute one of the biggest gene groups of multifunctional enzymes that assume essential parts in different physiological procedures, including assimilation, improvement and the resistant reaction (Zou et al. 2006). They are the foremost proteolytic stomach related catalysts in specific insects and hence give supplements required to survival and fecundity. All the known individuals from the chymotrypsin family have been found in creatures. It is striking that no individual from this exceptionally fruitful family has been experienced in protozoa or plants (Rawlings and Barrett 1993). SPs are synthesized as zymogens, which require proteolysis at a particular site for initiation. Enzymatically dynamic SPs include a high specificity reactant set of three amino acid residues in their synergist area, made of histidine (His), aspartic corrosive (Asp) and serine (Ser). Biochemical and genomic examinations uncovered that chemically inert serine protease homologs (SPH) are likewise individuals from the SP family (Zou et al. 2006). SPHs have comparative successions to SPs yet need at least one of the synergist build ups. Non-proteolytic SPHs are essential segments of phenoloxidase actuation in creepy crawly inborn resistant reactions (Yu et al. 2003). Bao et al. (2014) recognized an aggregate of 90 anticipated serine protease-like genes via seeking the *N. lugens* genome succession in view of the KEGG, Swissprot and Trembl comments, which were approved utilizing the tBLASTX calculation with a cut-off E-estimation of 10^{-10}. Serine proteinases can be classified into three groups based mainly on their primary substrate preference: (i) trypsin-like, which cleave after positively charged residue; (ii) chymotrypsin-like, which cleave after large hydrophobic residues; and (iii) elastase-like, which cleave after small hydrophobic residues.

Trypsin is the main intestinal digestive serine protease enzyme responsible for the hydrolysis of the peptide bonds in which the carboxyl groups are contributed by the lysine and arginine residues. Based on the ability of protease inhibitors to inhibit the enzyme from the insect gut, this enzyme has received attention as a target for biocontrol of insect pests. The enzyme is specifically inhibited by N-α-tosyl lysine chloromethyl ketone that acts on histidine (Omondi 2005). Through the use of ester or amide derivatives of arginine, such as N-α-tosyl arginine methyl ester (TAME) or N-α-Benzoyl-DL-arginine ethyl ester (BAEE) and N-α-Benzoyl-DL-arginine 4-nitroanilide (BApNA), digestive trypsin-like activity has been reported in most insect species (Applebaum 1985). Mostly, trypsin M_r values ranges from 20,000 to 35,000 Da and pI values are variable (range, 4–5). The pH optima is always alkaline (between 8 and 10), irrespective of the pH prevailing in midguts from which the trypsins are isolated. Nevertheless, trypsins isolated from Lepidoptera have higher pH optima that correspond to the higher pH values found in their midguts.

Isolation of inactive precursors (zymogens) of insect digestive proteinases has largely been unsuccessful (Applebaum 1985). Graf et al. (1986) suggested the occurrence of an inactive form of trypsin (trypsinogen) in midgut cells of *Aedes aegypti* (Stegomyia) based on the finding of trypsin immunoreactivity in midgut cells and on their failure to assay trypsin activity in homogenates of washed midgut cells. Nevertheless, trypsin is also immunolocalized in the glycocalyx of *Ae. aegypti* midgut cells (Graf et al. 1986), the site at which trypsin must be active and from where it cannot be removed by washing (Santos et al. 1986). The failure to assay trypsin in midgut homogenates indicates a low sensitivity of their assay procedure rather than in favour of the existence of a trypsinogen. Barillas-Mury et al. (1991) sequenced, what seemed to be a precursor of midgut trypsin in *Ae. aegypti*. Its sequence is similar to that of most trypsins, although it showed significant differences from the vertebrate trypsin precursors in the region of the activation peptide. Similar results were found with a putative trypsinogen from *Drosophila melanogaster* (Meigen) (Davis et al. 1985) and from *Simulium vittatum* (Zetterstedt) (Ramos et al. 1993). These differences suggest that the processing of precursors of insect trypsins may be different from that of vertebrates. There is evidence in *Tineola bisselliella* (Hummel) (Ward 1975) and *Bombyx mori* (Eguchi et al. 1982) that soluble trypsin is derived from membrane-bound forms. *Erinnyis ello* (Santos et al. 1986) and in *Musca domestica* (Linnaeus) (Lemos and Terra 1991a,b), trypsin is synthesized in midgut cells in an active form, but is associated with membranes of small vesicles. These vesicles then migrate to the cell apex and trypsin precursors are processed to a soluble form before being secreted. It seems that insects may control the activity of their digestive proteinases, in the absence of inactive forms, by binding the proteinases to membranes until they are released into the midgut lumen. Secretory granules isolated from the opaque zone cells from *Stomoxys calcitrans* (L.) adults contain a trypsin-like activity, which increases during incubation according to an apparent autocatalytic reaction (Moffatt and Lehane 1990). The finding by the authors that activation occurs to a different extent, if opaque zone cells are homogenized in the presence or absence of detergent, suggests that trypsin processing in this insect is also different from that found in vertebrates.

Serine proteases comprises a catalytic triad involving three amino acid residues, His, Asp and Ser, which are essential for the catalytic process. Formation of an acyl enzyme intermediate between the substrate and the Ser amino acid is the first step during the enzyme catalysis, which proceeds through a negatively charged tetrahedral transition state intermediate thereby resulting into the cleavage of the peptide bond. Ser-hydroxyl of the enzyme is restored during the second step, during which the acyl-enzyme intermediate is hydrolysed by a water molecule to release the peptide, hence the step is called deacylation reaction. The deacylation proceeds through the reverse reaction pathway of acylation which involves the formation of a tetrahedral transition state intermediate, where a water molecule is the attacking nucleophile instead of the Ser residue. The His residue provides a general base and accepts the OH group of the active Ser (Haq and Khan 2003; Haq et al. 2004).

Cysteine Proteinases

Cysteine proteinases, endopeptidyl hydrolases with a cysteine residue in their active center are generally recognized considering the impact of their active site inhibitors (iodoacetate, iodoacetamide and E-64) and activation of the catalysts by thiol compounds (Grudkowska and Zagdańska 2004). In insect pests, the cysteine proteinases are used in the digestive processes (Rawlings and Barrett 1993), however are also found in a few different tissues, showing that they may likewise assume different other roles (Matsumoto et al. 1998). pH reliance of cysteine proteinase movement in the unrefined concentrate of insect larvae have demonstrated that this action was for the most part in the basic range (Bode and Huber 1992; Oliveira et al. 2003). The papain family contains peptidases with a wide assortment of exercises, incorporating endopeptidases with wide specificity (for example, papain), endopeptidases with exceptionally limit specificity (for example, glycyl endopeptidases), aminopeptidases, dipeptidyl-peptidase, and peptidases with both endopeptidase and exopeptidase exercises (for example, cathepsins B and H). There are likewise relatives that demonstrate no reactant action (Dubey et al. 2007). All papain-like cysteine proteases share comparative successions (Berti and Storer 1995) and have comparative 3-D structures. The auxiliary information gives solid confirmation that these proteinases all emerged from a typical precursor (Dubey et al. 2007). Proteinaceous inhibitors of cysteine proteinases are subdivided into three families (stefin, cystatin and kininogen) considering their succession homology, the nearness and position of intrachain disulfide bonds, and the atomic mass of the protein (Turk and Bode 1991).

Cysteine proteases play an extensive variety of roles in insect pests, considering major functions in embryogenesis (Shiba et al. 2001), shedding (Liu et al. 2006), detoxification of plant protective proteins (Koo et al. 2008), insusceptible reactions (Zhang et al. 2013), and absorption (Goptar et al. 2012). The most widely contemplated part of cysteine proteases in herbivorous insect pests is their capacities as stomach related catalysts. In Coleoptera, Diptera, and Hemiptera, cysteine proteases are essential digestive enzymes (Cristofoletti et al. 2003). Numerous herbivorous insects utilize various sorts of proteases as digestive proteins. The exceptional assorted qualities and versatility of proteases expressible in the insects' nutritious tract empowers insect to safeguard themselves against an assortment of dietary poisons and antinutritional compounds they may experience in their host plants. Within the sight of protease inhibitors, insects can overproduce the current inhibitor-delicate stomach related proteases to surmount the inhibitors (Ahn et al. 2004) or increment articulation of inhibitor-inhumane protease isoforms (Bolter and Jongsma 1995; Oppert et al. 2010).

Cysteine proteinases catalyse the reaction in a similar way as serine proteinases through the formation of a covalent intermediate which involves a Cys and a His residue. The crucial Cys25 and His159 (e.g., papain) take part in the same role as Ser195 and His57, respectively, as in serine proteinases. Here the nucleophile is a thiolate ion instead of a hydroxyl group, which is stabilized through the formation of an ion pair with adjacent imidazolium group of His159. The attacking nucleophile is the thiolate-imidazolium ion pair in both steps, without the involvement of water molecule (Kuroda et al. 2001; Yoza et al. 2002; Connors et al. 2002; Haq et al. 2004).

Aspartic Proteinases

Insects have a wide range of proteases; the larger part utilizes serine proteases as essential digestive proteases (Waniek et al. 2005), and cysteine and aspartate proteases (cathepsins B, D, H, L) as intracellular lysosomal proteins (Cho et al. 1999). In some Coleoptera and cyclorrhaphous Diptera some portion of the midgut has an acidic pH of 5.4–6.9 and 3.1–6.8, separately (Terra and Ferreira 1994). In these insects, cysteine and aspartate proteases are emitted into the lumen of the midgut as significant digestive enzymes (Padilha et al. 2009). In the triatomine *Rhodnius prolixus* Stal, as per pH judgments by means of pH markers the pH esteem substitutes sustaining conditionally in the vicinity of 5.5 and 7.4 (J.M.C. Ribeiro and

E.S. Garcia, individual correspondence). Consequently, triatomines utilize those cathepsins as stomach related proteases (Terra et al. 1996).

Aspartic proteinases do not involve a covalent tetrahedral intermediate as observed in serine and cysteine proteinases. The nucleophilic attack is attained by two concurrent proton shifts: one from a water molecule to the diad of the two carboxyl groups and a second one from the diad to the carbonyl oxygen of the protein substrate with the concomitant peptide bond cleavage. This is a general acid-base catalysis, called a "push-pull" reaction mechanism, resulting to the formation of a non-covalent neutral tetrahedral intermediate (Mares et al. 1989).

Metalloproteinases

Metalloproteases are the most varied type of catalytic proteases characterized by the requirement for a divalent metal ion for their activity (Barrett 1998). They vary extensively in their amino acid sequences and their organization, and the great majority contain a zinc atom as their catalytically active site. Some cases involve another metal atom such as cobalt, manganese or nickel. Bacterial thermolysin is a well characterized metalloproteinase, and its crystallographic structure indicates that Zn is bound by two His and one Glu amino acid residues. Most of the metalloproteinases contain the amino acid sequence HEXXH, which provides two His ligands for the Zn binding while the third ligand is either a Glu (e.g., thermolysin, neprilysin, alanyl aminopeptidase) or a His (e.g., astacin). A water molecule is also essential for the catalysis, coordinates with the metal ion as a fourth ligand in the active form of the enzyme. Other families show a distinct mode of binding with the Zn atom. About 30 families of metalloproteases have been documented, of which 17 contain only endopeptidases, 12 contain only exopeptidases and 1 (M3) contains both endopeptidases and exopeptidases. An angiotensin-converting enzyme (ACE) in insects has substantially a functional metalloprotease with a presumed role in reproduction, development and defense (Macours and Hens 2004). Endothelin-converting enzyme (ECE) is another neuropeptide degrading metalloprotease reported from insects with an endopeptidase activity (Isaac 1988).

The mechanism of action of metalloproteases is vaguely different from that of other proteases in a way that they depend on the presence of bound divalent cation. The metal ion is held in position by several amino acid residues. The catalysis involves the formation of a non-covalent tetrahedral intermediate after the attack of a Zn-bound water molecule on the carbonyl group of the scissile peptide bond. This intermediate is finally decayed by transfer of proton from the Glu to the leaving group (Skiles et al. 2004).

Conclusion

The proteinases are a noteworthy group of hydrolytic enzymes in insects and are included in digestive processes. The voracious nature of the insect pests is mainly due to the presence of several isozymes of proteolytic enzymes in their gut. Characterization of the proteolytic properties of the digestive enzymes of insect pests therefore offer an opportunity for developing suitable and effective pest management strategies via plant protease inhibitors.

Insect Adaptations to Plant PIs

Defensive mechanisms against the insect pests developed in host plants pose a substantial selection pressure on them, which have resulted in development of counter adaptations to these defenses in insects (Gatehouse 2002; Jongsma and Bolter 1997). Although plants PIs are induced in response to insect damage, many insects have adapted to plant PIs resulting in even greater loss to the plants (Steppuhn and Baldwin 2007; Parde et al. 2012). This counter defense in insect pests in response to plant PIs is a major obstacle to the management of crop protection by exploitation of PIs for a long-lasting plant defense, and thus merits an understanding of the mechanisms by which insects counteract the PI-mediated plant defense. Adaptation mechanisms adapted to PIs in insect pests has attracted the researchers to understand the mechanisms, and eventually design better approaches so that PIs can be better utilized in crop protection (Parde et al. 2012; Bolter and Jongsma 1995; Brioschi et al. 2007). Insects overcome the insecticidal effect of plant PIs either by regulating the levels of existing proteases or by synthesizing newer proteases in their gut. Thus, there could be two types of resistance or adaptation mechanisms developed in insect pests in response to protease inhibitors. In one type, insects regulate the level of proteases in their midgut that are sensitive to the plant PIs or they may have mutations in gene encoding proteases which confer resistance without losing catalytic activity or over express the protease(s) that are insensitive to plant PIs (Parde et al. 2010). These insensitive proteases are produced either constitutively and/or induced in insects to compensate the loss of inhibited protease(s) (Parde et al. 2012; Bolter and Jongsma 1995; Jongsma et al. 1995). The second mechanism involves the alternative proteases, which either compensate the loss of PI-inhibited protease(s) or degrade the plant PIs to diminish their inhibitory activity (Zhu-Salzman et al. 2008; Giri et al. 1998) (Fig. 1). Hyper-secretion of additional proteinases in response to the inhibitors requires the utilization of essential amino acid pools that could starve the insects (Broadway and Duffey 1986; Broadway 1995). In addition, hyper-production of proteases in response to ingested

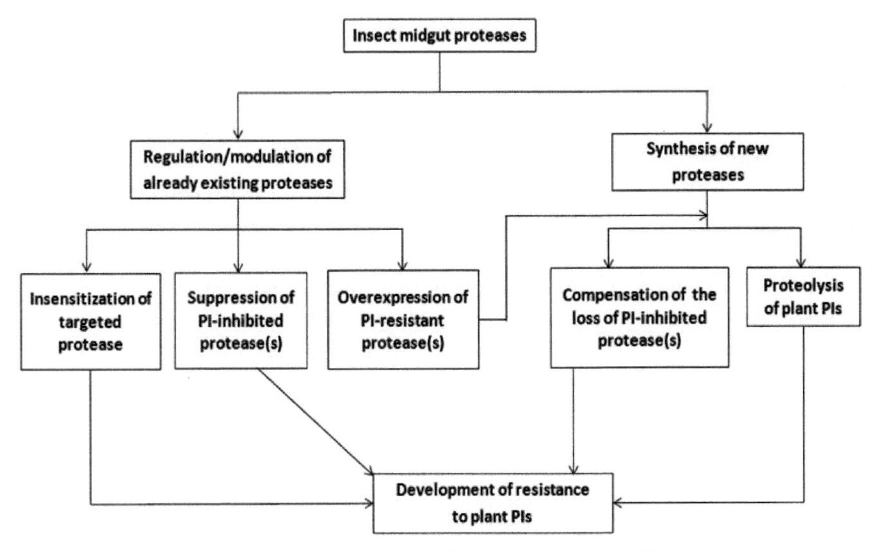

Figure 1. Adaptations of insect pests to plant PIs.

PIs leads to a further load on insect for energy and essential amino acids, resulting in delay of insect growth (Broadway and Duffy 1986). In contrast, few authors argue that the production of PI degrading proteinases derive dual benefit for insect by restoration of gut proteinase activity and the availability of valuable, sulfur-rich amino acids (Harsulkar et al. 1999).

H. armigera regulates its digestive proteinase levels against different types of PIs of *Albizia lebbeck* seeds by constitutive hyper-production of existing enzymes, trypsin, chymotrypsin and aminopeptidase activities to overcome the antinutritional effects of the inhibitor (Hivrale et al. 2013). Reduction in the serine protease activities due to ingestion of plant proteinase inhibitors is compensated with a significant induction of aminopeptidase activities in *Chilo suppressalis* and *Spodoptera exigua* (Lara et al. 2000; Vila et al. 2005). *H. armigera* larvae expressing high levels of chymotrypsin survive on a diet containing a multidomain serine PI from *Nicotiana alata* (Dunse et al. 2010). Naseri et al. (2010) demonstrated that larvae of *H. armigera* fed on soybean (cultivars L17 and Sahar) showed hyper-production of proteases in response to protease inhibition by PIs and leading to weak potential to increase its population on these cultivars. The inhibition of trypsin activity by PIs of these two soybean cultivars resulted in hyper-production of chymotrypsin-like enzymes in *H. armigera* (Naseri et al. 2010). Larvae reared on corn had the highest chymotrypsin- and elastase-like activity compared with other host plants to compensate the inhibitory effect of trypsin inhibitor of the host plant (Baghery et al. 2014). Wu et al. (1997) have reported the secretion of chymotrypsin- and elastase-like proteinases in *H. armigera* gut in response to giant taro trypsin inhibitor. This is because due to the broader substrate specificity and significant differential interaction of chymotrypsins with the inhibitors (Peterson et al. 1995). Within different host plants, the highest general proteolytic activity was in the larvae reared on cultivars Dehghan (white kidney bean) and Arman (chickpea), indicating the presence of some PIs on these cultivars, resulting in hyper-production of proteases by midgut cells of *H. armigera* in response to protease inhibition by PIs (Hemati et al. 2012).

Proteolytic inactivation is also a significant adaptation mechanism developed in insects to resist the proteolytic inhibition by PIs, wherein they have mutations which confer greater resistance without losing catalytic activity. Trypsins insensitive to plant PIs have been characterized from *Agrotis ipsilon, Trichoplusia ni* and *H. zea* (Volpicella et al. 2003; Broadway 1997; Mazumdar-Leighton and Broadway 2001). The larvae possessed higher levels of PI-resistant digestive proteolytic enzymes when fed on artificial diet incorporated with soybean trypsin inhibitor (SBTI) (Broadway 1997). Colorado potato beetles, *Leptinotarsa decemlineata* expressed cysteine proteinases resistant to inhibitors when fed on potato leaves containing high levels of endogenous proteinase inhibitors (Bolter and Jongsma 1995). Similarly, the expression of cysteine proteinases, intestains A and C, which are insensitive to the PIs, increased in Colorado potato beetle upon feeding on potato plants with induced PIs (Gruden et al. 2004). *Heliothis virescens* expressed PI-resistant trypsin enzyme when exposed to diet containing PIs (Jongsma et al. 1995; Gatehouse et al. 1997; Bown et al. 1997). *S. exigua* has developed resistance to potato proteinase inhibitor II by induced gut proteinase activity, which is insensitive to the inhibitors (Brioschi et al. 2007; Jongsma et al. 1995). Further, *S. frugiperda* when fed on diet containing Soybean Proteinase Inhibitors (SPI), the larval gut proteases were found to be insensitive to the inhibitor (Brioschi et al. 2007; Paulillo et al. 2000). A B-type carboxypeptidase in tomato fruitworm had developed resistance to the potato carboxypeptidase inhibitor due to the rearrangement of two small regions that otherwise stabilizes the enzyme-inhibitor complex resulting into a displacement of the active-site entrance, which impairs a proper interaction between the protease and its inhibitor (Bayes et al. 2005).

The regulation of synthesis of new enzymes resistant to the inhibitors is also one of the important adaptations in insects to plant PIs. Adaptation to SPI in *S. frugiperda* involves *de novo* synthesis and up-regulation of chymotrypsin and trypsin enzymes (Brioschi et al. 2007). A new trypsin-like protease is produced in *S. frugiperda* larvae when nurtured on artificial diet incorporated with soybean PIs (Paulillo et al. 2000). Some coleopteran and lepidopteran larvae exhibited proteolytic degradation of the PIs facilitated by the insect's midgut proteinases (Giri et al. 1998; Girard et al.

1998). The diamondback moth, *Plutella xylostella* larvae have been found to be insensitive to Mustard Trypsin Inhibitor 2 (MTI2), which has been attributed to the degradation of MTI2, thus preventing the inhibitory effect of the inhibitor (Yang et al. 2009).

It has been revealed that 12 different serine proteinases were either up- or down-regulated 2- to 12-fold in *H. armigera* when fed on soybean Kunitz-type trypsin inhibitor as evidenced by gene expression studies (Gatehouse et al. 1997; Bown et al. 1997). *Callosobruchus maculatus* counteracts soybean cysteine protease inhibitors (soyacystatin N, scN) by modulating digestive enzymes and about 30 different cDNAs encoding chief digestive cathepsin L-like cysteine proteases (CmCPs) have been copied (Zhu-Salzman et al. 2003). Based on sequence similarity these CmCPs can be CmCPA and CmCPB. CmCPB was over-expressed in bruchids when fed on diet containing scN, which has higher proteolytic activity, highly effective in converting zymogens into active forms and scN into inactive form (Ahn et al. 2004, 2007). The PIs, though considered as important and highly effective defense components of plant resistance, in most of the cases, no longer serve as resistant components in plants against insect pests.

Conclusions

The coevolution between plants and insects has lead to the development of important and effective plant defense systems in plants; however, insects too have developed several strategies to avoid plant defense systems. The counter defense by insects to plant defense is highly complex and has posed a big challenge in controlling them. The studies on insect adaptation have shown that even though plants develop highly effective and dynamic defensive strategies against insect pests, these strategies are vulnerable to insect adaptation in many ways. There is a need of in-depth studies on insect adaptations to plant defense to gain an understanding of the mechanisms underlying the adaptation, and the measures that need to be taken to prevent the insects from developing such adaptations.

References

Abdeen, A., A. Virgós, E. Olivella et al. 2005. Multiple insect resistance in transgenic tomato plants over-expressing two families of plant proteinase inhibitors. Plant Mol Biol. 57: 189–202.

Abe, K., Y. Emori, H. Kondo et al. 1987a. Molecular cloning of a cysteine proteinase inhibitor of rice (oryza cystatin). Homology with animal cystatins and transient expression in the ripening process of rice seeds. J Biol Chem. 262(35): 16793–16797.

Abe, K., Y. Emori, H. Kondo et al. 1988. The NH2-terminal 21 amino acid residues are not essential for the papain-inhibitory activity of oryzacystatin, a member of the cystatin superfamily. J Biol Chem. 263(16): 7655–7659.

Abe, M., H. Kondo and S. Arai. 1987b. Purification and characterization of a rice cysteine proteinase inhibitor. Agric Biol Chem. 51: 2763–2768.
Abe, M. and S. Arai. 1991. Some properties of a cysteine proteinase inhibitor from corn endosperm. Agric Biol Chem. 55(9): 2417–2418.
Abe, M., K. Abe, M. Kudora et al. 1992. Corn Kernel cysteine proteinase inhibitor as novel cystatin superfamily member of plant origin. Eur J Biochem. 209: 933–937.
Abrahanson, M., A. Ritonja, M.A. Brown et al. 1987. Identification of the probable inhibitory reactive sites of the cysteine proteinase inhibitors, human cystatin C and chicken cystatin. J Biol Chem. 262: 9688–9494.
Ahn, J.E., R.A. Salzman, S.C. Braunagel et al. 2004. Functional roles of specific bruchid protease isoforms in adaptation to a soybean protease inhibitor. Insect Mol Biol. 13: 649–657.
Ahn, J.E., M.R. Lovingshimer, R.A. Salzman et al. 2007. Cowpea bruchid Callosobruchusmaculatus counteracts dietary protease inhibitors through modulating properties of major digestive enzymes. Insect Mol Biol. 16: 295–304.
Akbar, S.M.D. and H.C. Sharma. 2017. Alkaline serine proteases from *Helicoverpa armigera*: potential candidates for industrial applications. Arch Insect Biochem Physiol. 94: e21367.
Al-Khunaizi, M., C.J. Luke, S. Cataltepe et al. 2002. The serpin SON-5 is a dual mechanistic class inhibitor of serine and cysteine proteinases. Biochemistry. 41(9): 3189–3199.
Altpeter, F., I. Diaz, H. McAuslane et al. 1999. Increased insect resistance in transgenic wheat stably expressing trypsin inhibitor CMe. Mol Breed. 5: 53–63.
Alvarez-Alfageme, F., J. Maharramov, L. Carrillo et al. 2011. Potential use of a serpin from Arabidopsis for pest control. PLoS One. 6: e20278.
Antcheva, N., A. Pintar, A. Patthy et al. 2001. Primary structure and specificity of a serine proteinase inhibitor from paprika (*Capsicum annuum*) seeds. Biochim Biophys Acta. 1298: 95–101.
Anwar, A. and M. Saleemuddin. 2002. Purification and characterization of a digestive alkaline protease from the larvae of *Spilosoma obliqua*. Arch. Insect Biochem Physiol. 51: 1–12.
Applebaum, S.W. 1985. Biochemistry of digestion. *In*: Kerkut, G.A. and L.I. Gilbert (eds.). Comprehensive Insect Physiology, Biochemistry, and Pharmacology. Pergamon Press, New York. 4: 279–311.
Arai, S., H. Watanabe, H. Kondo et al. 1991. Papain inhibitory activity of oryzacystatin, a rice seed cysteine proteinase inhibitor, depends on the central Gln-Val-Val-Ala-Gly region conserved among cystatin superfamily members. Biochem. 109: 294–298.
Arolas, J.L., J. Lorenzo, A. Rovira et al. 2005. A carboxypeptidase inhibitor from the Tick Rhipicephalus bursa, Isolation, cDNA cloning, recombinant expression and characterization. J Biol Chem. 280(5): 5441–5448.
Ascenzi, P., M. Ruoppolo, A. Amoresano et al. 1999. Characterization of low-molecular-mass trypsin isoinhibitors from oil-rape (*Brassica napus* var. oleifera) seed. Eur J Biochem. 261(1): 275–84.
Atiwetin, P., S. Harada and K. Kamei. 2006. Serine protease inhibitor from Wax Gourd (*Benincasa hispida* [Thumb] seeds). Biosci Biotech Biochem. 70(3): 743–745.
Atkinson, A.H., R.L. Heath, R.J. Simpson et al. 1993. Proteinase inhibitors in *Nicotiana alata* stigmas are derived from a precursor protein which is processed into five homologous inhibitors. Plant Cell. 5: 203–213.
Baghery, F., Y. Fathipour and B. Naseri. 2014. Digestive proteolytic and amylolytic activities in *Helicoverpa armigera* (Lep.: Noctuidae) larvae fed on five host plants. J Crop Prot. 3(2): 191–198.
Bao, Y.Y., X. Qin, B. Yu et al. 2014. Genomic insights into the serine protease gene family and expression profile analysis in the planthopper, *Nilaparvata lugens*. BMC Genomics. 15: 507.

Barillas-Mury, C., R. Graf, H.H. Hagedorn et al. 1991. cDNA and deduced amino acid sequence of a blood meal-induced trypsin from the mosquito, *Aedes aegypti*. Insect Biochem. 21: 825–831.

Barrett, A.J. 1987. The Cystatins, a new class of peptidase inhibitors. Trend Biochem Sci. 12: 193–196.

Barrett, A.J., N.D. Rawlings and J.F. Woessner (eds.). 1998. Handbook of Proteolytic Enzymes. Academic Press, London.

Bayes, A., M. Comellas-Bigler, M.R. de la Vega et al. 2005. Structural basis of the resistance of an insect carboxypeptidase to plant protease inhibitors. Proc Natl Acad Sci USA. 102(46): 16602–16607.

Belenghi, B., F. Acconcia, M. Trovato et al. 2003. Atcyssi, a cystatin from *Arabidopsis thaliana*, suppresses hypersensitive cell death. Eur J Biochem. 270: 2593–2604.

Bell, H.A., E.C. Fitches, R.E. Down et al. 2001. Effect of dietary cowpea trypsin inhibitor (CpTI) on the growth and development of the tomato moth *Lacanobia oleracea* (Lepidoptera: Noctuidae) and on the success of the gregarious ectoparasitoid *Eulophus pennicornis* (Hymenoptera: Eulophidae). Pest Manage Sci. 57: 57–65.

Berti, P.J. and A.C. Storer. 1995. Alignment/phylogeny of the papain superfamily of cysteine proteases. J Mol Biol. 246: 273–283.

Birk, Y., A. Gertler and S. Haled. 1963. A pure trypsin inhibitor from soya beans. Biochem J. 87: 281–284.

Birk, Y. 1985. The Bowman-Birk inhibitor: Trypsin and chymotrypsin inhibitor from soybeans. Int J Pept Protein Res. 25: 113–131.

Bode, W. and R. Huber. 1992. Natural protein proteinase inhibitors and their interaction with proteinases. Eur J Biochem. 204: 433–451.

Bolter, C.J. and M.A. Jongsma. 1995. Colorado potato beetles (*Leptinotarsa decemlineata*) adapt to proteinase-inhibitors induced in potato leaves by methyl jasmonate. J Insect Physiol. 41: 1071–1078.

Botella, M.A., Y. Xu., T.N. Prabha et al. 1996. Differentia expression of soybean cysteine proteinase inhibitor genes during development and in response to wounding and methyl jasmonate. Plant Physiol. 112: 1201–1210.

Boulter, D. 1993. Insect pest control by copying nature using genetically engineered crops. Biochem. 34: 1453–1466.

Bowman, D.E. 1946. Differentiation of soybean antitryptic factor. Proc Soc Exp Biol Med. 63: 547–550.

Bown, D.P., H.S. Wilkinson and J.A. Gatehouse. 1997. Differentially regulated inhibitor-sensitive and insensitive protease genes from the phytophagous insect pest, *Helicoverpa armigera*, are members of complex multigene families. Insect Biochem Mol Biol. 27: 625–638.

Breugelmans, B.G., Simonet, V. van Hoef and S. van Soest et al. 2009. Pacifastin-related peptides: structural and functional characteristics of a family of serine peptidase inhibitors. Peptides 30: 622–32.

Brioschi, D., L.D. Nadalini, M.H. Bengtsonb et al. 2007. General up regulation of Spodoptera frugiperda trypsins and chymotrypsins allows its adaptation to soybean proteinase inhibitor. Insect Biochem Mol Biol. 37: 1283–1290.

Broadway, R.M. and S.S. Duffey. 1986. Plant proteinase inhibitors: mechanism of action and effect on the growth and digestive physiology of larval *Heliothis zea* and *Spodoptera exigua*. J Insect Physiol. 32: 827–833.

Broadway, R.M. 1995. Are insects resistant to plant proteinase inhibitors? J Insect Physiol. 41: 107–116.

Broadway, R.M. 1997. Dietary regulation of serine proteinases that are resistant to serine proteinase inhibitors. J Insect Physiol. 43: 855–874.

Brovosky, D. 1986. Proteolytic enzymes and blood digestion in mosquito. Arch Insect Biochem Physiol. 3: 147–160.

Brown, W.M. and K.M. Dziegielewska. 1997. Friends and relations of the cystatin super family—new members and their evolution. Protein Sci. 6: 5–12.
Buchmanan, B.B., W. Gruissen and R.L. Jones. 2002. Biochemistry and molecular biology of plants. American Society of Plant Physiologists, Rockville, Maryland, USA, 1367p.
Cai, M., Y. Gong, J.L. Kao et al. 1995. Three-dimensional solution structure of Cucurbita maxima trypsin inhibitor-V determined by NMR spectroscopy. Biochemistry. 34(15): 5201–11.
Campos, R.G., J.A.T. Acosta, L.J.S. Aria et al. 1999. The use of cysteine proteinase inhibitors to engineer resistance against potyviruses in transgenic tobacco plants. Nat Biotechnol. 17: 1223–1226.
Ceci, L.R., N. Spoto, M. de Virgilio et al. 1995. The gene coding for the mustard trypsin inhibitor-2 is discontinuous and wound inducible. FEBS Lett. 364: 179–181.
Ceciliani, F., F. Bortolotti, E. Menengatti et al. 1994. Purification, inhibitory properties, amino acid sequence and identification of the reactive site of a new serine proteinase inhibitor from oil-rape (*Brassica napus*) seed. FEBS Lett. 342: 221–224.
Chapman, R.F. 1998. Alimentary canal, digestion and absorption. pp. 38–68. *In*: The Insects: Structure and Function. Cambridge: Cambridge Univ. Press. (4th ed.).
Charity, J.A., M.A. Anderson, D.J. Bittisnich et al. 1999. Transgenic tobacco and peas expressing a proteinase inhibitor from *Nicotiana alata* have increased insect resistance. Mol Breed. 5: 357–365.
Chen, P., J. Rose, R. Love et al. 1992. Reactive sites of an anticarcinogenic Bowman-Birk proteinase inhibitor are similar to other trypsin inhibitors. J Biol Chem. 267: 1990–1994.
Cho, W.L., S.M. Tsao, A.R. Hays et al. 1999. Mosquito cathepsin B-like protease involved in embryonic degradation of vitellin is produced as a latent extraovarian precursor. J Biol Chem. 274: 13311–13321.
Choi, D., J.A. Park, Y.S. Seo et al. 2000. Structure and stress-related expression of two cDNAs encoding proteinase inhibitor II of *Nicotiana glutinosa*. Biochim Biophys Acta. 1492(1): 211–215.
Chougule, N.P., E. Doyle, E. Fitches et al. 2008. Biochemical characterization of midgut digestive proteases from *Mamestra brassica* (cabbage moth: Lepidopter: Noctuidae) and effect of soybean Kunitz inhibitor (SKTI) in feeding assay. Journal of Insect Physiology. 54: 563–572.
Christeller, J. and W. Liang. 2005. Plant serine protease inhibitors. Prote in Peptide Lett. 12: 439–447.
Christy, L.A., S. Arvinth, M. Saravanakumar et al. 2009. Engineering sugarcane cultivars with bovine pancreatic trypsin inhibitor (aprotinin) gene for protection against top borer (*Scirpophaga excerptalis* Walker). Plant Cell Rep. 28: 175–184.
Colella, R., Y. Sakaguchi, H. Nagase et al. 1989. Chicken egg white cystatin. J Biol Chem. 264: 17164–17117.
Confalonieri, M., G. Allegro, A. Balestrazzi et al. 1998. Regeneration of *Populus Nigra* transgenic plants expressing a Kunitz Proteinase Inhibitor (Kti3) Gene. Mol Breed. 4: 137–145.
Connors, B.J., N.P. Laun, C.A. Maynard et al. 2002. Molecular characterization of a gene encoding a cystatin expressed in the stems of American chestnut (*Castanea dentata*). Planta. 215: 510–514.
Cowgill, S.E., C. Wright and H.J. Atkinson. 2002. Transgenic potatoes with enhanced levels of nematode resistance do not have altered susceptibility to non target aphids. Mol Ecol. 11: 821–827.
Cristofoletti, P.T., A.F. Ribeiro, C. Deraison et al. 2003. Midgut adaptation and digestive enzyme distribution in a phloem feeding insect, the pea aphid *Acyrthosiphon pisum*. J Insect Physiol. 49(1): 11–24. doi: 10. 1016/S0022-1910(02)00222-6. PMID:12770012.
Dahl, S.W., S.K. Rasmussen and J. Hejgaard. 1996. Heterologous expression of three plant serpins with distinct inhibitory specificities. J Biol Chem. 271: 25083–25088.

Davis, C.A., D.C. Riddell, M.J. Higgins et al. 1985. A gene family in *Drosophila melanogaster* coding for trypsin-like enzymes. Nucl Acid Res. 13: 6605–6619.

De Leo, F., M.A. Bonade-Bottino, L.R. Ceci et al. 1998. Opposite effects on *Spodoptera littoralis* larvae of high expression level of a trypsin proteinase inhibitor in transgenic plants. Plant Physiol. 118: 997–1004.

De Leo, F., L.R. Ceci, L. Jouanin et al. 2001. Analysis of mustard trypsin inhibitor-2 gene expression in response to developmental or environmental induction. Planta. 212: 710–717.

De Leo, F. and R. Gallerani. 2002. The mustard trypsin inhibitor 2 affects the fertility of *Spodoptera littoralis* larvae fed on transgenic plants. Insect Biochem Mol Biol. 32: 489–496.

Duan, X., X. Li, Q. Xue et al. 1996. Transgenic rice plants harboring an introduced potato proteinase inhibitor II gene are insect resistant. Nat Biotechnol. 14(4): 494–498.

Dubey, V.K., M. Pande, B.K. Singh et al. 2007. Papain-like proteases: Applications of their inhibitors. African J Biotechnol. 9: 1077–1086.

Dunaevskii, Y.E., I.P. Gladysheva, E.B. Pavlukova et al. 1997. The anionic protease inhibitor BBWI 1 from buckwheat seeds. Kinetic properties and possible biological role. Physiologia Plantarum. 100: 483–488.

Dunse, K.M., J.A. Stevens, F.T. Lay et al. 2010. Coexpression of potato type I and II proteinase inhibitors gives cotton plants protection against insect damage in the field. Proc Natl Acad Sci USA. 107: 15011–15015.

Dunse, K.M., Q. Kaas, R.F. Guarino et al. 2010. Molecular basis for the resistance of an insect chymotrypsin to a potato type II proteinase inhibitor. Proc Natl Acad Sci USA. 107: 15016–15021.

Eckelkamp, C., B. Ehmann and P. Schopfer. 1993. Wound-induced systemic accumulation of a transcript coding for a Bowman-Birk trypsin inhibitor-related protein in maize (*Zea mays* L.) seedlings. FEBS Lett. 323: 73–76.

Eguchi, M., A. Iwamoto and K. Yamauchi. 1982. Interrelation of proteases from the midgut lumen, epithelia and peritrophic membrane of the silkworm, *Bombyx mori*. Comp Biochem Physiol. 72(A): 359–363.

Falco, M.C. and M.C. Silva-Filho. 2003. Expression of soybean proteinase inhibitors in transgenic sugarcane plants: effects on natural defense against *Diatraea saccharalis*. Plant Physiol Biochem. 41: 761–766.

Felizmenio, Q.M.E., N.L. Daly and D.J. Craik. 2001. Circular proteins in plants: solution structure of a novel macrocyclic trypsin inhibitor from *Momordica cochinchinensis*. J Biol Chem. 276: 22875–22882.

Felton, G.W. and K.L. Korth. 2000. Trade-offs between pathogen and herbivore resistance. Curr Opin Plant Biol. 3: 309–314.

Fernandes, K.V.S., P.A. Sabelli, D.H.P. Barrett et al. 1993. The resistance of cow pea seeds to bruchid beetles is not related to level of cysteine proteinase inhibitors. Plant Mol Biol. 23: 215–219.

Ferry, N., L. Jouanin, L.R. Ceci et al. 2005. Impact of oilseed rape expressing the insecticidal serine protease inhibitor, mustard trypsin inhibitor-2 on the beneficial predator *Pterostichus madidus*. Mol Ecol. 14: 337–349.

Gatehouse, A.M., E. Norton, G.M. Davison et al. 1999. Digestive proteolytic activity in larvae of tomato moth, *Lacanobia oleracea*; effects of plant protease inhibitors *in vitro* and *in vivo*. J Insect Physiol. 45: 545–558.

Gatehouse, A.M.R. and D. Boutler. 1983. Assessment of the antimetabolic effects of trypsin inhibitors from cowpea (*Vigna unguiculata*) and other legumes on development of the bruchid beetle *Callosobruchus maculatus*. J Sci Food Agric. 34: 345–350.

Gatehouse, A.M.R., D. Boulter and V.A. Hilder. 1992. Potential of plantderived genes in the genetic manipulation of crops for insect resistance. pp. 155–181. *In*: Gatehouse, A.M.R., V.A. Hilder and D. Boulter (eds.). Plant Genetic Manipulation for Crop Protection. Redwood Press, Melcksham.

Gatehouse, A.M.R., G.M. Davison, C.A. Newell et al. 1997. Transgenic potato plants with enhanced resistance to the tomato moth, *Lacanobia oleracea*: growth room trials. Mol Breed. 3: 49–63.
Gatehouse, J.A. 2002. Plant resistance towards insect herbivores: a dynamic interaction. New Phytol. 156: 145–169.
Gatehouse, J.A. 2011. Prospects for using proteinase inhibitors to protect transgenic plants against attack by herbivorous insects. Curr Protein Pept Sci. 12: 409–16.
Gettins, P.G. 2002. Serpin structure, mechanism, and function. Chem Rev. 102: 4751–4804.
Gholizadeh, A., I.M. Santha, B.B. Kohnehrouz et al. 2005. Cystatins may confer viral resistance in plants by inhibition of a virus-induced cell death phenomenon in which cysteine proteinases are active: cloning and molecular characterization of a cDNA encoding cysteine proteinase inhibitor (celostatin) from celosia cristata (crested cock's comb). Biotechnol Appl Biochem. 42: 197204.
Girard, C., M. Le Metayer, M. Bonade-Bottino et al. 1998. High level of resistance to proteinase inhibitors may be conferred by proteolytic cleavage in beetle larvae. Insect Biochem Mol Biol. 28: 229–237.
Giri, A.P., A.M. Harsulkar, V.V. Deshpande et al. 1998. Chickpea defensive proteinase inhibitors can be inactivated by pod borer gut proteinases. Plant Physiol. 116: 393–401.
Gladysheva, I.P., T.S. Zamolodchikova, E.A. Sokolova et al. 1999. Interaction between duodenase, a proteinase with dual specificity, and soybean inhibitors of Bowman-Birk and Kunitz type. Biochemistry. 64: 1244–1249.
Goptar, I.A., T.A. Semashko, S.A. Danilenko et al. 2012. Cysteine digestive peptidases function as postglutamine cleaving enzymes in tenebrionid stored-product pests. Comp Biochem Physiol B Biochem Mol Biol. 161(2): 148–154. doi: 10.1016/j.cbpb. 2011.10.005. PMID: 22056682.
Gourinath, S., N. Alam, A. Srinivasan et al. 2000. Structure of the bifunctional inhibitor of trypsin and alpha-amylase from ragi seeds at 2.2 A resolution. Acta Crystallographica Section D. 56(3): 287–293.
Graf, R., A.S. Raikhel, M.R. Brown et al. 1986. Mosquito trypsin: immunocyto-chemical localization in the mid-gut of blood-fed *Aedes aegypti* (L.). Cell Tiss Res. 245: 19–27.
Graham, J.S. and C.A. Ryan. 1997. Accumulation of metallocarboxypeptidase inhibitor in leaves of wounded potato plants. Biochem Biophys Res Comm. 101: 1164–1170.
Gruden, K., A.G. Kuipers, G. Guncar et al. 2004. Molecular basis of Colorado potato beetle adaptation to potato plant defence at the level of digestive cysteine proteinases. Insect Biochem Mol Biol. 34(4): 365–75.
Grudkowska, M. and B. Zagdańska. 2004. Multifunctional role of plant cysteine proteinases. Acta Biochim Polonica. 51: 609–624.
Grzonka, Z., E. Jankowska, R. Kasprzykowska et al. 2001. Structural studies of cysteine proteases and their inhibitiors. Acta Biochem Pol. 48: 1–20.
Habib, H. and K.M. Fazili. 2007. Plant protease inhibitors: a defense strategy in plants. Biotechnology and Molecular Biology Review. 2(3): 068–085.
Haq, S.K. and R.H. Khan. 2003. Characterization of a proteinase inhibitor from *Cajanus cajan* L. J Protein Chem. 22: 543–554.
Haq, S.K., S.M. Atif and R.H. Khan. 2004. Protein proteinase inhibitor genes in combat against insects, pests, and pathogens: natural and engineered phytoprotection. Arch Biochem Biophys. 431: 145–159.
Hara, S., J. Makino and T. Ikenaka. 1989. Amino acid sequences and disulfide bridges of serine proteinase inhibitors from bitter gourd (*Momordica charantia* Linn) seeds. J Biochem. 105: 88–92.
Harsulkar, A.M., A.P. Giri, A.G. Patankar et al. 1999. Successive use of non-host plant proteinase inhibitors required for effective inhibition of *Helicoverpa armigera* gut proteinases and larval growth. Plant Physiol. 121(2): 497–506.
Hass, G.M., H. Nau, K. Biemann et al. 1975. The amino acid sequence of a carboxypeptidase inhibitor from potatoes. Biochem. 14: 14–142.

Hass, G.M. and M.A. Hermodson. 1981. Amino acid sequence of a carboxypeptidase inhibitor from tomato fruit. Biochemistry. 20: 2256–2260.
Havkioja, E. and L. Neuvonen. 1985. Induced long-term resistance to birch foliage against defoliators: defense or incidental. Ecology. 66: 1303–1308.
Heath, R.L., G. McDonald, J.T. Christeller et al. 1997. Proteinase inhibitors from *Nicotiana alata* enhance plant resistance to insect pests. J Insect Physiol. 43: 833–842.
Heitz, A., J.F. Hernandez, J. Gagnon et al. 2001. Solution structure of the squash trypsin inhibitor MCOTI-II. A new family for cyclic knottins. Biochemistry. 40: 7973–7983.
Hemati, S.A., B. Naseri, G.N. Ganbalani et al. 2012. Digestive proteolytic and amylolytic activities and feeding responses of *Helicoverpa armigera* (Lepidoptera: Noctuidae) on different host plants. J Econ Entomol. 105(4): 1439–1446.
Hilder, V.A., A.M. Gatehouse, S.E. Sheerman et al. 1987. A novel mechanism of insect resistance engineered into tobacco. Nature. 300: 160–163.
Hines, M.E., C.I. Osuala and S.S. Nielsen. 1991. Isolation and partial characterization of a soybean cystatin cysteine proteinase inhibitor of Coleopteran digestive proteolytic activity. J Agric Food Chem. 39: 1515–1520.
Hivrale, V.K., P.R. Lomate, S.S. Basaiyye et al. 2013. Compensatory proteolytic responses to dietary proteinase inhibitors from *Albizialebbeck* seeds in the *Helicoverpa armigera* larvae. Arthropod Plant Interact. 7: 259–266.
Hoffmann, M.P., F.G. Zalom, L.T. Wilson et al. 1992. Field evaluation of transgenic tobacco containing genes encoding *Bacillus thuringiensis* ∂-endotoxin or cowpea trypsin inhibitor: efficacy against *Helicoverpa zea* (Lepidoptera: Noctuidae). J Econ Entomol. 85: 2516–2522.
Hollander-Czytko, H., J.L. Andersen and C.A. Ryan. 1985. Vacuolar localization of wound-induced carboxy peptidase inhibitor in potato leaves. Plant Physiol. 78: 76–79.
Homandberg, G.A., R.D. Litwiller and R.J. Peanasky. 1989. Carboxypeptidase inhibitors from *Ascaris suum*: the primary structure. Arch Biochem Biophys. 270: 153–161.
Hoorn, R.A.L. and D.G. Jones. 2004. The plant proteolytic machiney and its role in defense. Curr Opin Plant Biol. 7: 400–407.
Hosoyama, H., K. Irie, K. Abe et al. 1994. Oryzacystatin exogenously introduced into proplasts and regeneration of transgenic rice. Biosci Biotechnol Biochem. 58: 1500–1505.
Houseman, J.G. and A.E.R. Downe. 1983. Cathepsin D-like activity in the posterior midgut of Hemipteran insects. Comp Biochem Physiol-B. 75: 509–512.
Houseman, J.G., A.E.R. Downe and B.J.R. Philogene. 1989. Partial characterization of proteinase activity in the larval midgut of the European corn borer *Ostrinia nubilalis* Hubner (Lepidoptera: Pyralidae). Can J Zool. 67: 864–868.
Huang, Y.J., K.Y. To, M.N. Yap et al. 2001. Cloning and characterization of leaf senescence up-regulated genes in sweet potato. Physiol Plant. 113: 384–391.
Huntington, J.A., R.J. Read and R.W. Carrell. 2000. Structure of serpin protease complex shows inhibition by deformation. Nature. 407(6806): 923–926.
Irie, K. 1996. Transgenic rice established to express corn cystatin exhibits strong inhibitory activity against insect gut proteinases. Plant Mol Biol. 30: 149–157.
Irving, J.A., P.J. Steenbakkers, A.M. Lesk et al. 2002a. Serpins in prokaryotes. Mol Biol Evol. 19(11): 1881–1890.
Irving, J.A., R.N. Pike, W. Dai et al. 2002b. Evidence that serpin architecture intrinsically supports papain-like cysteine protease inhibition: engineering alpha (I) antitrypsin to inhibit cathepsin proteases. Biochemistry. 41: 4998–5004.
Irving, J.A., S.S. Shushanov, R.N. Pike et al. 2002c. Inhibitory activity of a heterochromatin-associated serpin (MENT) against papain like cysteine proteinases affects chromatin structure and blocks cell proliferation. J Biol Chem. 277: 13192–13201.
Isaac, R.E. 1988. Neuropeptide degrading activity of locust (Schistocerca gregaria) synaptic membranes. Biochem J. 255: 843–847.
Ishihara, M., K. Atta, S. Tawata et al. 1999. Purification and characterization of intracellular cysteine protease inhibitor from *Chlorella* sp. Food Sci Technol Res. 5: 210–213.

Ishihara, M., M. Morine, T. Tiara et al. 2000. Purification and characterization of intracellular cysteine protease inhibitor from *Chlorella* sp. Food Sci Technol Res. 6: 161–165.

Ishikawa, A., S. Ohta, K. Matsuoka et al. 1994. A family of potato genes that encode Kunitz type proteinase inhibitors: structural comparisons and differential expression. Plant Cell Physiol. 35: 303–312.

Iwasaki, T., T. Kiyohara and M. Yoshikawa. 1971. Purification and partial characterization of two different types of proteinase inhibitors (inhibitors II-a and II-b) from potatoes. J Biochem. 70: 817–826.

Johnson, R., J. Narvaez and G. An. 1989. Expression of proteinase inhibitors I and II in transgenic tobacco plants: effects on natural defense against *Manduca sexta* larvae. Proc Natl Acad Sci USA. 86: 9871–9875.

Jongsma, M.A., P.L. Bakker, J. Peters et al. 1995. Adaptation of *Spodoptera exigua* larvae to plant proteinase-inhibitors by induction of gut proteinase activity insensitive to inhibition. Proc Natl Acad Sci USA. 92: 8041–8045.

Jongsma, M.A. and C. Bolter. 1997. The adaptation of insects to plant protease inhibitors. J Insect Physiol. 43: 885–895.

Jongsma, M.A. and J. Beekwilder. 2011. Co-evolution of insect proteases and plant protease inhibitors. Curr Protein Pept Sci. 12: 437–47.

Joshi, B., M. Sainani, K. Bastawade et al. 1998. Cysteine protease inhibitor from pearl millet: a new class of antifungal protein. Biochem Biophys Res Comm. 246: 382–387.

Keilova, H. and V. Tomasek. 1976. Isolation and properties of cathepsin D inhibitor from potatoes. Collect Czech Chem Commun. 41: 489–497.

Kenny, A.J. 1999. Introduction: Nomenclature and classes of peptidases. pp. 1–8. *In*: E.E. Sterchi and W. Stocker (eds.). Proteolytic Enzymes. Tools and Targets. Berlin: SpringerVerlag.

Kervinen, J., G.J. Tobin, J. Costa et al. 1999. Crystal structure of plant aspartic proteinase prophytepsin: inactivation and vacuolar targeting. J Eur Mol Biol Organ. 18(14): 3947–3955.

Kim, J.Y., Y.S. Chung, K.H. Paek et al. 1999. Isolation and characterization of a cDNA encoding the cysteine proteinase inhibitor, induced upon flower maturation in carnation using suppression subtractive hybridization. Mol Cells. 9(4): 392–397.

Koo, Y.D., J.E. Ahn, R.A. Salzman et al. 2008. Functional expression of an insect cathepsin B-like counter-defence protein. Insect Mol Biol. 17(3): 235–245. doi:10.1111/j.1365-2583.2008.00799.x. PMID: 18397276.

Kortt, A.A. Strike and P.M. De Jersey. 1989. Amino acid sequence of a crystalline seed albumin (Winged Bean Albumin-l) from *Psophocarpus tetraffonolobus* (L.) DC . Sequence similarity with Kunitz-type seed inhibitors and 7S storage globulins. Eur J Biochem. 181: 403–408.

Kuntiz, M. 1945. Crystallization of a trypsin inhibitor from soybean. Science. 101: 668–669.

Kuroda, M., T. Kiyosaki, I. Matsumoto et al. 2001. Molecular cloning, characterization and expression of wheat cystatins. Biosci Biotechnol Biochem. 65: 22–28.

Laing, W.A. and M.T. McManus. 2002. Proteinase inhibitors. *In*: McManus, M.T., W.A. Laing and A.C. Allan (eds.). Protein-protein Interactions in Plants. Sheffield Academic Press, UK. 7: 77–119.

Lara, P., F. Ortego, E. Gonzalez-Hidalgo et al. 2000. Adaptation of *Spodoptera exigua* (Lepidoptera: Noctuidae) to barley trypsin inhibitor BTI-CMe expressed in transgenic tobacco. Transgenic Res. 9: 169–178.

Laskowski, M. and I. Kato. 1980. Protein inhibitors of proteinases. Annu Rev Biochem. 49: 593–626.

Law, R., Q. Zhang, S. McGowan et al. 2006. An overview of the serpin superfamily. Genome Biol. 7(5): 216.

Lawrence, P.K. and K.R. Koundal. 2002. Plant protease inhibitors in control of phytophagous insects. Electron J Biotechnol. 5(1): 93–109.

Ledoigt, G., B. Griffaut, E. Debiton et al. 2006. Analysis of secreted protease inhibitors after water stress in potato tubers. Int J Biol Macromols. 38: 268–271.

Lee, C.F. and J.Y. Lin. 1995. Amino acid sequences of trypsin inhibitors from the melon Cucumis melo. J Biochem. 118(1): 18–22.

Lee, J.S., W.E. Brown, J.S. Graham et al. 1986. Molecular characterization and phylogenetic studies of a wound-inducible proteinase inhibitor I gene in (*Lycopersicon* species). PNAS. 83(19): 7277–7281.

Lemos, F.J.A. and W.R. Terra. 1991a. Properties and intracellular distribution of a cathepsin D-like proteinase active at the acid region of *Musca domestica* mid-gut. Insect Biochem. 21: 457–465.

Lemos, F.J.A. and W.R. Terra. 1991b. Digestion of bacteria and the role of mid-gut lysozyme in some insect larvae. Comp Biochem Physiol. 100(B): 265–268.

Ligoxygakis, P., S. Roth and J.M. Reichhart. 2003. A serpin regulates dorsal ventral axis formation of the Drosophila embryo. Curr Biol. 13: 2097–2102.

Lin, G.D., W. Bode, R. Huber et al. 1993. The 0.25 nm X-ray structure of the Bowman-Birk type inhibitor from mung bean in ternary complex with porcine trypsin. Eur J Biochem. 212: 549–555.

Lin, Y.H., H.T. Li, Y.C. Huang et al. 2006. Purification, crystallization and preliminary X-ray crystallographic analysis of rice Bowman-Birk inhibitor from *Oryza sativa.* Acta Cryst. F62: 522–524.

Lipker, H., G.S. Fraenkel and I.E. Liener. 1954. Effect of soybean inhibitors on growth of *Tribolium confusum.* J Sci Food Agric. 2: 410–415.

Liu, H.B., X. Guo, M.S. Naeem et al. 2011. Transgenic *Brassica napus* L. lines carrying a two gene construct demonstrate enhanced resistance against *Plutella xylostella* and *Sclerotinia sclerotiorum.* Plant Cell Tiss Org Cult. 106: 143–151.

Liu, J., G.P. Shi, W.Q. Zhang et al. 2006. Cathepsin L function in insect moulting: molecular cloning and functional analysis in cotton bollworm, *Helicoverpa armigera.* Insect Mol Biol. 15(6): 823–834.

Liu, Q., L. Yu, J. Gao et al. 2000. Cloning, tissue expression pattern and genomic organization of latexin, a human homologue of rat carboxypeptidase A inhibitor. Mol Biol Rep. 27: 241–246.

Luo, M., Z. Wang, H. Li et al. 2009. Overexpression of a weed (*Solanum americanum*) proteinase inhibitor in transgenic tobacco results in increased glandular trichome density and enhanced resistance to *Helicoverpa armigera* and *Spodoptera litura.* Int J Mol Sci. 10: 1896–1910.

Macedo, M.L.R., F.R. Caio, de Oliveira, M. Poliene et al. 2015. Adaptive mechanisms of insect pests against plant protease inhibitors and future prospects related to crop protection: A review. Protein Peptide Lett. 22: 149–163.

Machleidt, W., U. Borchart, H. Fritz et al. 1983. Protein inhibitors of cysteine proteinase II. Primary structure of stefin a cytosolic inhibitor of cysteine proteinases from human polymorphonuclear granulocytes. Hoppe-Seyler's Physiol Chem. 364: 1481–1486.

Macours, N. and K. Hens. 2004. Zinc-metalloproteases in insects: ACE and ECE. Insect Biochem Mol Biol. 34: 501–510.

Maheswaran, G., L. Pridmore, P. Franz et al. 2007. A proteinase inhibitor from *Nicotiana alata* inhibits the normal development of light-brown apple moth, *Epiphyas postvittana* in transgenic apple plants. Plant Cell Rep. 26: 773–782.

Mahoney, W.C., M.A. Hermodson, B. Jones et al. 1984. Amino acid sequence and secondary structural analysis of the corn inhibitor of trypsin and activated Hageman factor. J Biol Chem. 259: 8412–8416.

Marchetti, S., M. Delledonne, C. Fogher et al. 2000. Soybean Kunitz, C-II and PI-IV inhibitor genes confer different levels of insect resistance to tobacco and potato transgenic plants. Theor Appl Genet. 101: 519–526.

Mares, M., B. Meloun, M. Pavlik et al. 1989. Primary structure of cathepsin D inhibitor from potatoes and its structure relationship to soybean trypsin inhibitor family. FEBS Lett. 251: 94–98.
Margossian, L.J., A.D. Federman, J.J. Giovannoni et al. 1988. Ethylene-regulated expression of a tomato fruit ripening gene encoding a proteinase inhibitor I with a glutamic residue at the reactive site. Proc Natl Acad Sci USA. 85(21): 8012–8016.
Matsumoto, I., K. Abe, S. Arai et al. 1998. Functional expression and enzymatic properties of two *Sitophilus zeamais* cysteine proteinases showing different autolytic processing profiles *in vitro*. J Biochem. 123: 693–700.
Mazumdar-Leighton, S. and R.M. Broadway. 2001. Transcriptional induction of diverse midgut trypsins in larval Agrotisipsilon and *Helicoverpa zea* feeding on the soybean trypsin inhibitor. Insect Biochem Mol Biol. 31(6-7): 645–57.
McManus, M., E.P.J. Burgess, B. Philip et al. 1999. Expression of the soybean (Kunitz) trypsin inhibitor in transgenic tobacco: Effects on larval development of *Spodoptera litura*. Transgenic Res. 8: 383–395.
McManus, M.T., D.W.R. White and P.G. McGregor. 1994. Accumulation of a chymotrypsin inhibitor in transgenic tobacco can affect the growth of insect pests. Transgenic Res. 3: 50–58.
McWherter, C.A., W.F. Malkenhorst, J. Campbell et al. 1989. Novel inhibitors of human leukocyte elastase and cathepsin G: sequence variants of squash seed protease inhibitor with altered protease selectivity. Biochem. 28: 5708–5714.
Menengatti, E., G. Tedeschi, S. Ronchi et al. 1992. Purification, inhibitory properties and amino acid sequence of a new serine proteinase inhibitor from white mustard (*Sinapis alba* L.) seed. FEBS Lett. 301: 10–14.
Mickel, C.E. and J. Standish. 1947. University of Minnesota Agricultural Experimental Station Technical Bulletin. pp. 178, 1–20.
Miller, E.A., M.C. Lee, A.H. Atkinson et al. 2000. Identification of a novel four-domain member of the proteinase inhibitor II family from the stigmas of *Nicotiana alata*. Plant Mol Biol. 42(2): 329–333.
Moffatt, M.R. and M.J. Lehane. 1990. Trypsin is stored as an inactive zymogen in the mid-gut of *Stomoxys ealcitrans*. Insect Biochem. 20: 719–723.
Mosolov, V.V. and T.A. Valueva. 2008. Proteinase inhibitors in plant biotechnology: a review. Appl Biochem Microbiol. 44: 233–40.
Mukhopadhyay, D. 2000. The molecular evolutionary history of a winged bean α-chymotrypsin inhibitor and modeling of its mutations through structural analyses. J Mol Evol. 50: 214–223.
Murdock, L.L., G. Brookhart, P.E. Dunn et al. 1987. Cysteine digestive proteinases in Coleoptera. Comp Biochem Physiol. 87B: 783–787.
Murray, C. and J.T. Christeller. 1995. Purification of a trypsin inhibitor (PFTI) from pumpkin fruit phloem exudate and isolation of putative trypsin and chymotrypsin inhibitor cDNA clones. Biol Chem Hoppe-Seyler. 376(5): 281–287.
Nandi, A.K., D. Basu, S. Das et al. 1999. High level expression of soybean trypsin inhibitor gene in transgenic tobacco plants failed to confer resistance against damage caused by *Helicoverpa armigera*. J Biosci. 24: 445–452.
Naseri, B., Y. Fathipour, S. Moharramipour et al. 2010. Digestive proteolytic and amylolytic activities of *Helicoverpa armigera* in response to feeding on different soybean cultivars. Pest Manag Sci. 66: 1316–1323.
Neurath, H. 1984. Evolution of proteolytic enzymes. Science. 224: 350–357.
Normant, E., M.P. Martres, J.C. Schwartz et al. 1995. Purification, cDNA cloning, functional expression, and characterization of a 26 kDa endogenous mammalian carboxypeptidase inhibitor. Proc Natl Acad Sci USA. 92: 12225–12229.
Odani, S., T. Koide and T. Ono. 1983. The complete amino acid sequence of barley trypsin inhibitor. J Biol Chem. 258: 7998–8003.

Odani, S., T. Koide and T. Ono. 1986. Wheat gram trypsin inhibitor. Isolation and structural characterization of single-headed and double-headed inhibitors of the Bowman-Birk type. J Biochem. 100(4): 975–983.

Oliveira, A.S., J.X. Filho and M.P. Sales. 2003. Cysteine proteinases cystatins. Braz Arch Biol Technol. 46(1): 91–104.

Omondi, J.G. 2005. Digestive endo-proteases from the mid-gut glands of the Indian white shrimp, *Penaeus indicus* (Decapoda: Penaeidae) from Kenya. Western Indian Ocean Journal. 4(1): 109–121.

Oppert, B., E.N. Elpidina, M. Toutges et al. 2010. Microarray analysis reveals strategies of *Tribolium castaneum* larvae to compensate for cysteine and serine protease inhibitors. Comp Biochem Physiol D. 5(4): 280–287. doi: 10.1016/j.cbd.2010.08.001. PMID: 20855237.

Otto, H.H. and T. Schirmeister. 1997. Cysteine proteases and their inhibitors. Chem Rev. 97: 133–171.

Padilha, M.H., A.C. Pimentel, A.F. Ribeiro et al. 2009. Sequence and function of lysosomal and digestive cathepsin D-like proteinases of *Musca domestica* midgut enzymes. Insect Biochem Mol Biol. 39: 782–791.

Parde, V.D., H.C. Sharma and M.S. Kachole. 2010. *In vivo* inhibition of *Helicoverpa armigera* gut pro-proteinase activation by non host plant protease inhibitors. J Insect Physiol. 56: 1315–1324.

Parde, V.D., H.C. Sharma and M.S. Kachole. 2012. Potential of protease inhibitors in wild relatives of pigeonpea against the cotton bollworm/legume pod borer, *Helicoverpa armigera*. Am J Plant Sci. 3: 627–635.

Park, H., N. Yamanaka, A. Mikkonem et al. 2000. Purification and characterization of aspartic proteinase from sunflower seeds. Biosci Biotechnol Biochem. 64: 931–999.

Park, Y., B.H. Choi, J.S. Kwak et al. 2005. Kunitz-type serine protease inhibitor from potato (*Solanum tuberosum* L. cv. Jopung). J Agric Food Chem. 53: 6491–6496.

Paulillo, L.C.M.S., A.R. Lopes, P.T. Cristofoletti et al. 2000. Changes in midgut endopeptidase activity of Spodopterafrugiperda (Lepidoptera: Noctuidae) are responsible for adaptation to soybean proteinase inhibitors. J Econ Entomol. 93: 892–896.

Pearce, G., S. Johnson and C.A. Ryan. 1993. Purification and characterization from Tobacco Nicotiana tabacum) leaves of six small, wound-inducible, proteinase isoinhibitors of the potato inhibitor II family plant physiol. 102: 639–644.

Pernas, M., M.R. Sanchez, L. Gomez et al. 1998. A chestnut seed cystatin differentially effective against cysteine proteinases from closely related pests. Plant Mol Biol. 38: 1235–1242.

Pernas, M., R. Sanchez-Mong and G. Salcedo. 2000. Biotic and abiotic stress can induce cystatin expression in chestnut. FEBS Lett. 467(2-3): 206–210.

Peterson, A.M., G.J.P. Fernando and M.A. Wells. 1995. Purification, characterization and cDNA sequence of an alkaline chymotrypsin from the midgut of *Manduca sexta*. Insect Biochem Mol Biol. 25: 765–774.

Plunkett, G., D.F. Senear, G. Zuroske et al. 1982. Proteinase inhibitors I and II from leaves of wounded tomato plants: purification and properties. Arch Biochem Biophys. 213: 463–472.

Prakash, B., S. Selvaraj, M.R.N. Murthy et al. 1996. Analysis of the aminoacid sequences of plant Bowman-Birk inhibitors. J Mol Evol. 42: 560–569.

Pyati, P., A.R. Bandani, E. Fitches et al. 2011. Protein digestion in cereal aphids (*Sitobion avenae*) as a target for plant defense by endogenous proteinase inhibitors. J Insect Physiol. 57: 881–91.

Qi, R.F., Z. Song and C. Chi. 2005. Structural features and molecular evolution of Bowman-Birk protease inhibitors and their potential application. Acta Biochim Biophys Sin (Shanghai). 37(5): 283–292.

Rahbe, Y., C. Deraison, M. Bonade-Bottino et al. 2003. Effects of the cysteine protease inhibitor oryzacystatin on different expression (OC-I)n different aphids and reduced

performance of Myzus persicae on OC-I expressing transgenic oilseed rape. Plant Sci. 164: 441–450.
Raj, S.S., E. Kibushi, T. Kurasawa et al. 2002. Crystal structure of bovine trypsin and wheat germ trypsin inhibitor (I-2b) complex (2:1) at 2.3 a resolution. J Biochem. 132: 927–933.
Ramos, A., A. Mahowald and M. Jacobs-Lorena. 1993. Gut-specific genes from the black-fly *Simulium vittatum* encoding trypsin-like and carboxy peptidase-like proteins. Insect Molec Biol. 1: 149–163.
Rancour, J.M. and C.A. Ryan. 1968. Isolation of a carboxypeptidase B inhibitor from potatoes. Arch Biochem Biophys. 125: 380–382.
Rawlings, N.D. and A.J. Barrett. 1993. Evolutionary families of peptidases. Biochem J. 290: 205–218.
Rawlings, N.D., D.P. Tolle and A.J. Barrett. 2004. Evolutionary families of peptidase inhibitors. J Biochem. 378: 705–716.
Ray, C.A., R.A. Black, S.R. Kronheim et al. 1992. Viral inhibition of inflammation, cow pox virus encodes an inhibitor of interleukin-1beta converting enzyme. Cell. 69: 597–604.
Reeck, G.R., K.J. Kramer, J.E. Baker et al. 1997. Proteinase inhibitors and resistance of transgenic plants to insects. pp. 157–183. *In*: Carozzi, N. and M. Koziel (eds.). Advances in Insect Control: The Role of Transgenic Plants. London, Taylor and Francis.
Reverter, D., J. Vendrell, F. Canals et al. 1998. A carboxypeptidase inhibitor from the medical leech Hirudo medicinalis. Isolation, sequence analysis, cDNA cloning, recombinant expression, and characterization. J Biol Chem. 273: 32927–32933.
Richardson, M.J. 1991. Seed storage proteins: Academic Press, 259–305.
Ritonja, A., I. Krizaj, P. Mesko et al. 1990. The amino acid sequence of a novel inhibitor of cathepsin D from potato. FEBS Lett. 267: 1–15.
Roberts, T.H., S. Marttila, S.K. Rasmussen et al. 2003. Differential gene expression for suicide-substrate serine proteinase inhibitors (serpins) in vegetative and grain tissues of barley. J Exp Bot. 54: 2251–2263.
Roger, L.M., E.S. Hart, S.B. Kaye et al. 2000. Bean a-amylase inhibitor 1 in transgenic peas (*Pisumsativum*) provides complete protection from pea weevil (*Bruchus pisorum*) under field conditions. Proc Natl Acad Sci USA. 97(8): 3820–3825.
Rolka, K., G. Kupryszewski, J. Rozycki et al. 1992. New analogues of *Cucurbita maxima* trypsin inhibitor III (CMTI III) with simplified structure. Biol Chem Hoppe-Seyler. 373: 1055–1060.
Ryan, C.A. and A.K. Balls. 1962. An inhibitor of chymotrypsin from *Solanum tuberosm* and its behavior toward trypsin. Proc Natl Acad Sci USA. 48: 1839–44.
Ryan, C.A. 1990. Proteinase inhibitors in plants: genes for improving defenses against insects and pathogens. Annu Rev Phytopathol. 28: 425–449.
Ryan, S.N., W.A. Liang and M.T. McManus. 1998. A cysteine proteinase inhibitor purified from apple fruit. Phytochemistry 49: 957–63.
Sakuta, C., A. Konishi, S. Yamakawa et al. 2001. Cysteine proteinase gene expression in the endosperm of germinating carrot seeds. Biosci Biotechnol Biochem. 65: 2243–2248.
Salvesen, G., C. Parkes, M. Abrahanson et al. 1986. Human low-Mr kininogen contain three copies of a cystatin sequence that are divergent in structure and in inhibitory activity for cystine proteinases. Biochem J. 234: 429–434.
Sanatan, P.T., P.R. Lomate, A.P. Giri et al. 2013. Characterization of a chemostable serine alkaline protease from *Periplaneta americana*. BMC Biochemistry. 14: 32.
Sane, V., P. Nath and S.P. Aminuddin. 1997. Development of insect-resistant transgenic plants using plant genes: Expression of cowpea trypsin inhibitor in transgenic tobacco plants. Curr Sci. 72: 741–747.
Santos, C.D., A.F. Ribeiro and W.R. Terra. 1986. Differential centrifugation, calcium precipitation and ultrasonic disruption of midgut cells of *Erinnyis ello* caterpillars. Purification of cell microvilli and inferences concerning secretory mechanisms. Can J Zool. 64: 490–500.

Schechter, I. and A. Berger. 1967. On the size of the active site in proteases. I. Papain. Biochem Biophys Res Commun. 27: 157–62.

Senthilkumar, R., C.P. Cheng and K.W. Yeh. 2010. Genetically pyramiding protease-inhibitor genes for dual broad-spectrum resistance against insect and phytopathogens in transgenic tobacco. Plant Biotechnol J. 8: 65–75.

Sharma, K. 2015. Protease inhibitors in crop protection from insects. Int J Curr Res and Academic Review. 3(2): 55–70.

Shiba, H., D. Uchida, H. Kobayashi et al. 2001. Involvement of cathepsin B- and L-like proteinases in silk gland histolysis during metamorphosis of *Bombyx mori*. Arch Biochem Biophys. 390(1): 28–34. doi:10.1006/abbi.2001. 2343. PMID: 11368511.

Shivraj, B. and T.N. Pattabiraman. 1981. Natural plant enzyme inhibitors. Characterization of an unusual alpha-amylase/trypsin inhibitor from ragi (*Eleusine coracana* Geartn.). Biochem J. 193: 29–36.

Shulke, R.H. and L.L. Murdock. 1983. Lipoxygenase trypsin inhibitor and lectin from soybeans: effects on larval growth of *Manduca sexta* (Lepidoptera: Sphingidae). Env Ento. 12: 787–791.

Skiles, J.W., N.C. Gonnella and A.Y. Jeng. 2004. The design, structure, and clinical update of small molecular weight matrix metalloproteinase inhibitors. Curr Med Chem. 11: 2911–2977.

Solomon, M., B. Belenghi, M. Delledonne et al. 1999. The involvement of cysteine protease inhibitor genes in the regulation of programmed cell death in plants. Plant Cell. 11: 431–443.

Spit, J., B. Breugelmans, V. van Hoef et al. 2012. Growth-inhibition effects of pacifastin-like peptides on a pest insect: the desert locust, *Schistocerca gregaria*. Peptides. 34: 251–57.

Srinivasan, A., A.P. Giri and V.S. Gupta. 2006. Structural and functional diversities in lepidopteran serine proteases. Cell Mol Biol Lett. 11: 132–154.

Stato, N., K. Ishidoh, E. Uchiyama et al. 1990. Molecular cloning and sequencing of cDNA from rat ystatin B. Nucleic Acids Res. 18(22): 6698.

Steller, H. 1995. Mechanisms and genes of cellular suicide. Science. 267: 1445–1449.

Steppuhn, A. and I.T. Baldwin. 2007. Resistance management in a native plant: nicotine prevents herbivores from compensating for plant protease inhibitors. Ecol Lett. 10: 499–511.

Strobl, S., K. Maskos, G. Wiegand et al. 1998. A novel strategy for inhibition of alphaamylases: yellow meal worm alpha-amylase in complex with the Ragi bifunctional inhibitor at 2.5 A resolution. Structure. 6: 911–921.

Tanaka, A.S., M.U. Sampaio, S. Marangoni et al. 1997. Purification and primary structure determination of a Bowman-Birk trypsin inhibitor from *Torresea cearensis* seeds. Biol Chem. 378: 273–281.

Tashiro, M., K. Hashino, M. Shiozaki et al. 1987. The complete aminoacid sequence of rice bran trypsin inhibitor. J Biochem. 102: 297–306.

Terra, W.R. 1990. Evolution of digestive systems of insects. Annu Rev Entomol. 35: 181–200.

Terra, W.R. and C. Ferreiri. 1994. Insect digestive enzymes—Properties, compartmentalization and function. Comp Biochem Physiol B-Biochem Mol Biol. 109: 1–62.

Terra, W.R., C. Ferreira and J.E. Baker. 1996. Compartimentalization of digestion. pp. 206–235. *In*: Lehane, M.J. and P.F. Billingsley (eds.). Biology of the Insect Midgut. Chapman & Hall, London.

Terra, W.R., C. Ferreira and B.P. Jordao. 1996. Digestive enzymes. pp. 153–194. *In*: Lehane M.J. (ed.). Billin London, Chapman and Hall.

Thaimattam, R., E. Tykarska, A. Bierzynski et al. 2002. Atomic resolution structure of squash trypsin inhibitor: unexpected metal coordination. Acta Crystallogr D Biol Crystallogr. 58: 1448–1461.

Tur, S.A., Y. Birk, A. Gertler et al. 1972. A basic trypsin and chymotrypsin inhibitor from groundnuts (*Arachis hypogaea*). Biochim Biophys Acta. 263: 666–672.

Turk, B., V. Turk and D. Turk. 1997. Structural and functional aspects of papain like cysteine proteinase and their protein inhibitors. Biol Chem Hoppe-Seyler. 378: 141–150.
Turk, V. and W. Bode. 1991. The cystatins: protein inhibitors of cysteine proteinases. FEBS Lett. 285: 213–219.
Ussuf, K.K., N.H. Laxmi and R. Mitra. 2001. Proteinase inhibitors. Plant derived genes of insecticidal protein for developing insect-resistant transgenic plants. Current Sci. 80, 7(10): 47–853.
Valueva, T.A., T.A. Revina, E.L. Gvozdeva et al. 2001. Effects of elicitors on the accumulation of proteinase inhibitors in injured potato tubers. Appl Biochem Microbiol. 37: 512–516.
Van derVyver, C., J. Schneidereit, S. Driscoll et al. 2003. Oryzacystatin I expression in transformed tobacco produces a conditional growth phenotype and enhances chilling tolerance. Plant Biotech J. 1(2): 101–112.
Vierstra, R.D. 1996. Proteolysis in plants: mechanisms and functions. Plant Mol Biol. 32: 275–302.
Vila, L., J. Quilis, D. Meynard et al. 2005. Expression of the maize proteinase inhibitor (*mpi*) gene in rice plants enhances resistance against the striped stem borer (*Chilo suppressalis*): effects on larval growth and insect gut proteinases. Plant Biotechnol J. 3: 187–202.
Volpicella, M., A. Schipper, M.A. Jongsma et al. 2000. Characterization of recombinant mustard trypsin inhibitor 2 (MTI2) expressed in Pichia pastoris. FEBS Lett. 468(2-3): 137–41.
Volpicella, M., L.R. Ceci, J. Cordewener et al. 2003. Properties of purified gut trypsin from *Helicoverpa zea*, adapted to proteinase inhibitors. Eur J Biochem. 270: 10–19.
Walker, A.J., L. Ford, M.E.N. Majerus et al. 1998. Characterisation of the midgut digestive proteinase activity of the two-spot ladybird (*Adalia bipunctata* L.) and its sensitivity to proteinase inhibitors. Insect Biochemistry and Molecular Biology. 28: 173–180.
Walsh, T.A. and J.A. Strictland. 1993. Proteolysis of the 85 kDa crystalline cysteine proteinase inhibitor from tomato release functional cystatin domains. Plant Physiol. 103(4): 1227–1234.
Wang, H.X. and T.B. Ng. 2006. Concurrent isolation of a Kunitz-type trypsin inhibitor with antifungal activity and a novel lectin from *Pseudostellaria heterophylla* roots. BBRC. 342(1): 349–353.
Waniek, P.J., U.B. Hendgen-Cotta, P. StockMayer et al. 2005. Serine proteinases of the human body louse (Pediculus humanus): sequence characterization and expression patterns. Parasitol Res. 97: 486–500.
Ward, C.W. 1975. Resolution of proteases in the keratinolytic larvae of the webbing clothes moth. Aust J biol Sci. 28: 1–23.
Watanabe, H., K. Abe, Y. Emori et al. 1991. Molecular cloning and gibbrllin-induced expression of multiple cystine protease of rice seds (orizains). J Biol Chem. 266: 16897–16902.
Wieczorek, M., J. Otlewski, J. Cook et al. 1985. The squash family of serine proteinase inhibitors. Amino acid sequences and association equilibrium constants of inhibitors from squash, summer squash, zucchini, and cucumber seeds. Biochem Biophys Res Commun. 126: 646–652.
Williamson, V.M. and R.S. Hussey. 1996. Nematode pathogenesis and resistance in plants. Plant Cell. 8(10): 35–1745.
Wingate, V.P., R.M. Broadway and C.A. Ryan. 1989. Isolation and characterization of a novel, developmentally regulated proteinase inhibitor I protein and cDNA from the fruit of a wild species of tomato. J Biol Chem. 264(30): 17734–17738.
Winterer, J. and J. Bergelson. 2001. Diamondback moth compensatory consumption of protease inhibitor transformed plants. Mol Ecol. 10: 1069–1074.
Wolfson, J.L. and L.L. Murdock. 1987. Suppression of larval Colorado beetle growth and development by digestive proteinase inhibitors. Entomologia Experimentalis et Applicata. 44: 235–240.

Wu, Y., D. Llewellyn, A. Mathews et al. 1997. Adaptation of *Helicoverpa armigera* (Lepidoptera: Noctuidae) to a proteinase inhibitor expressed in transgenic tobacco. Mol Breed. 3: 371–380.

Xu, D., Q. Xue, D. McElroy et al. 1996. Constitutive expression of a cowpea trypsin inhibitor gene, CpTi, in transgenic rice plants confers resistance to two major rice insect pests. Mol Breed. 2: 167–173.

Yang, L., Z. Fang, M. Dicke et al. 2009. The diamondback moth, *Plutella xylostella*, specifically inactivates Mustard Trypsin Inhibitor 2 (MTI2) to overcome host plant defence. Insect Biochem Mol Biol. 33: 55–61.

Yeh, K.W., M.I. Lin, S.J. Tuan et al. 1997. Sweet potato (*Ipomoea batatas*) trypsin inhibitors expressed in transgenic tobacco plants confer resistance against *Spodoptera litura*. Plant Cell Rep. 16: 696–699.

Yoo, B.C., K. Aoki, Y. Xiang et al. 2000. Characterization of *Cucurbita maxima* phloem serpin-1 (CmPS-1). A developmentally regulated elastase inhibitor. J Biol Chem. 275: 35122–35128.

Yoza, K., S. Nakamura, M. Yaguchi et al. 2002. Molecular cloning and functional expression of cDNA encoding a cysteine proteinase inhibitor, cystatin, from Job's tears (Coix lacryma-jobi L. var. Mayuen Stapf). Biosci Biotechnol Biochem. 66: 2287–2291.

Yu, X.-Q., H. Jiang, Y. Wang et al. 2003. Nonproteolytic serine proteinase homologs are involved in prophenoloxidase activation in the tobacco hornworm, Manduca sexta. Insect Biochem Mol Biol. 33(2): 197–208.

Zhang, Y., Y.X. Lu, J. Liu et al. 2013. A regulatory pathway, ecdysone-transcription factor relish-cathepsin L, is involved in insect fat body dissociation. PLOS Genet. 9(2): 1003273.

Zhao, Y., M.A. Botella, L. Subramanian et al. 1996. Two wound-inducible soybean cysteine proteinase inhibitors have greater insect digestive proteinase inhibitory activities than a constitutive homolog. Plant Physiol. 111(4): 1299–306.

Zhu-Salzman, K., H. Koiwa, R.A. Salzman et al. 2003. Cowpea Bruchid Callosobruchusmaculatus uses a three-component strategy to overcome a plant defensive cysteine protease inhibitor. Insect Mol Biol. 12: 135–145.

Zhu-Salzman, K., D.S. Luthe and G.W. Felton. 2008. Arthropod-inducible proteins: Broad spectrum defenses against multiple herbivores. Plant Physiol. 146: 852–858.

Zou, Z., D.L. Lopez, M.R. Kanost et al. 2006. Comparative analysis of serine protease-related genes in the honey bee genome: possible involvement in embryonic development and innate immunity. Insect Mol Biol. 15(5): 603–614.

CHAPTER 2

Genetic, Biochemical and Molecular Networks of Plant-Insect Interactions

Model Platforms for Integrative Biological Research

Jessica Lasher, Allison Speer, Samantha Taylor and *Chandrakanth Emani**

INTRODUCTION

Plant-insect interactions in nature showcase unique genetic, biochemical and molecular network patterns that encompass different levels of biological organization (Schoonhoven et al. 2005), from subcellular mechanisms at the cytological and genetic levels to biochemical and molecular functions (Kessler and Baldwin 2002) at the ecological community level (Bezemer and van Dam 2005; Kessler and Halitschke 2007). In the current scenario of systems biology research, the study of plant-insect interactions thus provides model research platforms for researchers to conduct integrative research in genetics, biochemistry, molecular biology and more importantly, in the field of ecology in an agronomic background (Snoeren et al. 2007).

Induced Defence in Plant-Insect Interactions

Among the plant pests, insects are the most diverse and abundant group and about 45% of the approximately 1 million described insect species (Stork 2007) are herbivores (Schoonhoven et al. 2005). The insect attack on plants was shown to be below ground as well as above ground (Bezemer and van Dam 2005), with every plant organ being equally targeted resulting

Department of Biology, Western Kentucky University-Owensboro, 4821 New Hartford Road, Owensboro, KY 42303.
* Corresponding author: chandrakanth.emani@wku.edu

in an array of induced defenses. The myriad biochemical and molecular mechanisms that plants evolved to tackle the diverse insect herbivores (Walling 2000) include both constitutive and induced defense mechanisms (Schoonhoven et al. 2005). The diversity of plant herbivores also makes it impossible for plants to evolve constitutive defenses against all potential attackers. Further, constitutive defenses may act as a natural selection mechanism for adaptations that occur in herbivorous insects (Agrawal and Karban 1999). On the other hand, inducible defenses are more favored as they are tuned to the needs of the specific plants and depend on the actual presence of the insect herbivores and have the potential to modify the plant phenotype while limiting the adaptations in herbivores (Agrawal and Karban 1999). A specific illustration is the herbivore-induced biochemical production in an attacked plant that suppresses a direct defense mechanism while having no effect on a putative indirect defense against an adapted herbivore (Kahl et al. 2000). Specifically, *Nicotiana attenuata* plants respond to insect attack by induction of nicotine that is a neurotoxin that affects most insect herbivores (Kahl et al. 2000). However, the same plant has a highly-weakened nicotine induction in response to feeding damage by caterpillars of the specialist tobacco hornworm *Manduca sexta* that can tolerate high levels of nicotine (Krischik et al. 1988), and a more interesting observation was the induction of volatile terpenoids that can attract parasitic wasps that attack the caterpillars (Kahl et al. 2000).

Molecular analyses of plant responses to insect herbivory showed that induced defenses comprise direct defenses, such as secondary metabolites and protease inhibitors that negatively affect herbivore growth and survival (Kessler and Baldwin 2002), as well as indirect defenses in the form of plant volatiles that mediate interactions between plants and arthropods (D'Alessandro and Turlings 2006). A specific example for both direct and indirect chemical defence against folivorous insect pests was seen in pine plants (Mumm and Hilker 2006). An evolutionary change from induced to constitutive expression of indirect plant resistance was observed (Heil et al. 2004). *In vitro* studies comprising the genetic engineering of terpenoid metabolism in the model plant *Arabidopsis thalina* showed indirect induced effects such as the attraction of plants to carnivorous predatory mites (*Phytoseiulus persimilis*) that further the plants' defense mechanisms (Kappers et al. 2005). The induced defence mechanisms are mostly specific for the attacker species (Kahl et al. 2000; Dicke et al. 2003; Arimura et al. 2005) thus enabling an individual plant to express a range of different phenotypes where each phenotype has its own effects on diverse members of the community, such as herbivores, carnivores, and pollinators (Dicke et al. 2004; Kessler et al. 2004; Kessler and Halitschke 2007; Bruinsma and Dicke 2008). A herbivore-induced plant vaccination is observed in nature where a previous attack by a herbivore results in the induction of specific subsequent defenses that significantly affect herbivore and plant fitness (Kessler and Baldwin 2004; Voelckel and Baldwin 2004b). This necessitates a comprehensive investigation of induced defenses that takes into account the total community as well as individual interactions among community members (Kessler and Baldwin 2004; Bruinsma and Dicke 2008), specifically as seen in the studies of silencing of the jasmonate cascade involved in plant-insect interactions (Kessler et al. 2004). Studies that combined transcript and metabolite analysis revealed an array of genes involved in insect herbivore induced volatile formation in plants thus pointing to mechanisms related to transcriptional induction, metabolite induction, and ecological interaction (Mercke et al. 2004). This proves that integrative approaches at different levels of biological organization may be a major challenge for researchers uncovering the molecular mechanisms involved in plant-insect interactions. Identification of genes involved in specific plant-insect interactions are always followed by in-depth studies investigating their ecological function where specific genes of interest have been silenced (Paschold et al. 2007). This enables the investigation of gene function at different levels of biological integration providing novel insights into the ecological function of specific genes.

The Signal Transduction Networks

A typical manifestation of induced defenses in plants in response to an insect herbivore attack is in the form of a phytohormone-mediated signal transduction that links the damage caused by the attack to specific phenotypic changes in the plant. The well-researched signal transduction pathways that form part of the induced defenses are (a) the jasmonate pathway related to the phytohormone jasmonic acid (JA), (b) the shikimate pathway of salicylic acid (JA), and (c) the ethylene pathway related to ethylene (ET) (Dicke and Van Poecke 2002; Kessler and Baldwin 2002).

The most crucial pathway among the signal transduction networks as seen in both insect-induced and wound-induced plant responses is the jasmonate pathway where the JAZ or Jasmonate ZIM (Zinc Finger Inflorescence Meristem) domain repressor proteins target the jasmonate signaling pathway direct and indirect responses (Thines et al. 2007). Specific instances of jasmonate induction was observed in tobacco by *Manduca sexta* caterpillars (Kahl et al. 2000), by *Tetranychus urticae* (spider mite) in tomato through the octadecanoid signaling pathway (Li et al. 2002), and by *Pieris rapae* caterpillars or *Frankliniella occidentalis* thrips in *Arabidopsis thaliana* (Reymond et al. 2004; De Vos et al. 2005). Jasmonate-inducible defence mechanisms result in increased parasitism of insect herbivores (Thaler 1999) and in inducing direct as well as indirect defenses in plants at the molecular level (Kessler and Baldwin 2002). Specifically, in conserved transcript patterns, 114 insect responsive genes were identified in *Arabidopsis thaliana* induced by *Pieris rapae* including genes involved in pathogenesis, indole glucosinolate metabolism, detoxification and cell survival along with the jasmonate signal transduction pathway genes (Reymond et al. 2004).

The other significant induced response apart from JA in plants, involves the SA related responses as seen in the phloem-feeding silver leaf whitefly *Bemisia tabaci* that actually suppressed JA-dependent defenses while inducing the SA-dependent defenses (Kempema et al. 2007). This observation suggests that individual insect herbivores elicit distinct phytohormone signatures accompanied by specific transcriptome changes resulting in dynamic phytohormone induction patterns (De Vos et al. 2005).

Insect Herbivory Induced Global Transcriptome Changes

The individual plant response to insect herbivores is usually exhibited in the form of extensive rearrangements of gene transcription involving a diverse array up- or down-regulated genes. This was illustrated through microarray analysis of SA and JA-induced studies in tobacco (Heidel and Baldwin 2004), conserved transcript patterns in Arabidopsis (Reymond et al. 2004), lepidopteran larval induction studies in tobacco (Voelckel and Baldwin 2004a), signaling pathway analysis in Arabidopsis (Bodenhausen and Reymond 2007), genome-wide transcription profiling in *Brassica oleracea* (Broekgaarden et al. 2007) and transcriptome changes induced by phloem-feeding insects in Arabidopsis (Kempema et al. 2007). This provides extensive new information about the first steps of plant responses to insect attack. These studies established that there appears to a vast array of mechanisms involved in the transcriptomic response of a plant to diverse insect herbivores (Voelckel and Baldwin 2004a; De Vos et al. 2005) as well as responses of different plant cultivars in response to the same herbivore (Broekgaarden et al. 2007). In *Arabidopsis thaliana* exposed to cell content-feeding thrips (*Frankliniella occidentalis*) and tissue-biting caterpillars (*Pieris rapae*) induced similar numbers of differentially expressed genes. However, the gene sets differentially responding to these two herbivore species were completely different in that only 9%–17% of the JA-responsive genes exhibited similar responses to *F. occidentalis* and *P. rapae* (De Vos et al. 2005). Another interesting study in Arabidopsis involving the induction of responses by the phloem-feeding silverleaf white fly nymphs (*B. tabaci*) compared to aphids (*Myzus persicae*) showed completely different corresponding transcriptomic changes (Kempema et al. 2007).

Genotypic variation in plants can also elicit different transcriptional responses to the same insect herbivore as seen from the differences in genome-wide transcription profiles in two white cabbage (*Brassica oleracea var. capitata*) cultivars in response to *Pieris rapae* caterpillar attack (Broekgaarden et al. 2007). The parallel profiling of genome-wide transcriptomic changes with metabolite production and expression of different levels of resistance throws novel insights into the plant-insect interactions in varied ecological backgrounds.

Mechanical wounding and insect feeding induced completely different transcript profiles in Arabidopsis as seen in a study that examined cDNA microarray profiles of about 150 genes involved in tissue defense and repair (Reymond et al. 2000). The wound-induced genes that were also regulated by water stress were induced either to lesser levels or were completely absent when compared to larval feeding, thus suggesting larval feeding strategies may minimize the activation of a subset of water stress-inducible defense-related genes (Reymond et al. 2000). In contrast, similar expression patterns of conserved transcripts were observed in Arabidopsis leaves in response to feeding by both specialist *P. rapae* and generalist *Spodoptera littoralis* caterpillars (Reymond et al. 2004). The transcript profiles observed in signaling pathways controlling induced resistance were different in ET and SA mutants (Bodenhausen and Reymond 2007). Native tobacco plants, *Nicotiana attenuata*, elicited different transcriptional responses thus distinguishing between attacks by two generalist polyphagous larvae (*Spodoptera exigua* and *Heliothis virescens*) and a specialist oliphagous larva (*M. sexta*) (Voelckel and Baldwin 2004a). A large overlap in transcriptional responses was observed in the cDNA microarray enriched in defence-related genes for all three lepidopteran larvae, but the plants responded more similarly to attack from the two generalists than to attack from the specialist. The fatty acid-amino acid conjugate profiles of the polyphagous *S. exigua* and *H. virescens* larvae were almost identical, but those from the oligophagous *M. sexta* larvae were different in being dominated by fatty acid-Glu conjugates (Voelckel and Baldwin 2004a). Quantitative real-time polymerase chain reaction (PCR) studies enabled the detailed expression analysis of temporal, spatial and density-dependent dynamics of transcriptomic changes for a limited number of genes of particular interest such as the lipoxygenase pathway genes involved in early steps of plant responses to insect herbivores (Zheng et al. 2007). Transcriptome and metabolome changes were also examined at a time series-based investigation that enabled proposing a model of plant-insect herbivore interactions at early phase of infestation (Kusnierczyk et al. 2008). These research advances enable agronomists and ecologists to gather valuable insights into the timing and dynamics of early responses of plants to insect herbivores.

Herbivore Induced Plant Vaccination

A plant can be conditioned to be more resistant by a process akin to vaccination as an initial attack from herbivores or pathogens may result in subsequent attacks from the same species, and may also affect different species (Kessler and Baldwin 2004). This cross resistance termed immunization or vaccination can confer a natural benefit to the plant if the arising fitness consequences of the initial herbivore attack are less than subsequent herbivore attacks (Kessler and Baldwin 2004). This is a most probable scenario in an agronomic setting as plants are seldom attacked by a single herbivore species and the more general and likely scenario would be multiple attacks by different herbivores and pathogens in a sequential or simultaneous process. In nature, the occurrence of plant-mediated competition between herbivorous insects and between insects and pathogens is also a most likely scenario observed by ecologists (Kaplan and Denno 2007). However, limited investigations exist that throw light on the mechanisms underlying the effects of defenses induced by an initial insect herbivore attack on other members of the community. More recent studies can now establish a clear interspecific relation between the ecological phenomena of community effects of plant infestation with mechanical information. Studies in *Arachis hypogaea* (peanut) plants revealed that initial infection with the white mold fungus *Sclerotium rolfsii* make them preferable for subsequent infestation by beet armyworm moths (*Spodoptera exigua*) for oviposition. However, a sequence of events now result in beet armyworm moth larval feeding on fungus-infected plants to enhanced risks of attack by parasitic wasps. Consequently, volatiles emitted from plants infested with fungus in combination with the caterpillars were more effective in conferring resistance than those with volatiles emitted from when infested only with caterpillars (Cardoza et al. 2003).

It was observed that Arabidopsis plants attacked by the caterpillar *P. rapae* were found develop resistance to several other plant pathogens other than insect pests including Turnip crinkle virus (De Vos et al. 2006). This was surprising because resistance to Turnip crinkle virus is dependent on Salicyclic acid (SA) signal transduction pathway, while *P. rapae* specifically induces jasmonic acid (JA) and ethylene (ET) pathways but not the SA pathway. This led the researchers to conclude that caterpillar infestation induced ET primed the Arabidopsis plants for augmented expression of SA-dependent defense mechanisms (De Vos et al. 2006). Tomato plants damaged by aphids (*Macrosiphum euphorbiae*) were preferred for oviposition by beet armyworm moths and their larvae fed in increasing number on aphid-infested plants than on uninfested control plants (Rodriguez-Saona et al. 2005). When cultivated tomato plants were treated with the phytohormone JA during the early planting season in the field, they were found to have long-lasting effects on community dynamics under field conditions (Thaler 1999).

Other examples of plant vaccination involved multiple attacks of insect herbivores on plants that were found to induce indirect defenses. *P. rapae* infestation were found to indirectly affect reduced paratization of *Plutella xylostella* caterpillars by the parasitoid *Cotesia plutellae* as it was found to result in a reduced attraction to plants infested by both caterpillar species through increased volatile production as compared to infestation by *P. xylostella* alone. In a related scenario, another interesting observation is of the preferential oviposition of the adult females of *P. xylostella* on plants already infested by *P. rapae* caterpillars (Shiojiri et al. 2002).

Nicotina attenuata (tobacco) plants that are initially attacked by the mirid bug *Tupiocoris notatus* were observed to have increased resistance to *Manduca sexta* caterpillars and these plants upon transcriptomic and metabolite analysis were found to have modified direct and indirect defence pathways (Kessler and Baldwin 2004; Voelckel and Baldwin 2004b). All these research studies that are illustrations of integrative approaches that led to plant vaccination in nature due to initial attack by insect herbivores pointed to a coordinated levels of biological integration and threw novel insights in a comprehensive understanding of multiple plant-attacker interactions.

A more interesting phenomenon that was arose out of the plant vaccination observations was that of an "evolutionary arms race" between the plants and their corresponding insect herbivores (Schoonhoven et al. 2005). An illustration of this phenomenon was what was termed as "molecular sabotage" of plant defense pathways when it was observed that herbivores may inject their salivary components in the plant they attack to overcome plant defence pathways (Will et al. 2007). In the resulting arms race scenario, plants would exploit the same salivary components of herbivores to induce defence pathways (Kessler and Baldwin 2002). As detailed earlier in the chapter, since plant defense responses are regulated by a network of signal transduction pathways, it can now be concluded that the arms race phenomenon results as an integrative cross talk between the various signal transduction pathways. In a specific illustration, the SA and JA pathways can negatively interact during the induction of defence mechanisms, while the JA and ET pathways often show positive interaction (Pieterse et al. 2002; Koornneef and Pieterse 2008). The resulting integrative network signal transduction pathways may benefit the plants in terms of improved resistance by modifying their defenses against a diverse array of insect herbivores in the field. This led to the conclusion of jasmonate and salicylate metabolites christened as global signals for improved defense gene expression (Reymond and Farmer 1998) and this was clearly illustrated by the herbivore-induced resist against diverse microbial pathogens in Arabidopsis (De Vos et al. 2006). This beneficial integrative phenomenon also had a negative connotation as it provided insect herbivores with an ability to interfere with the induction of defence pathways. An illustration of this observation was in Arabidopsis (Zarate et al. 2007) in a study investigating the basal defenses related to the resistance to the phloem-feeding silverleaf whitefly (SLWF; *Bemisia tabaci*). When sentinel defense gene RNAs were monitored in SLWF-infested and control plants, it was found that Salicylic acid (SA)-responsive gene transcripts (*PR1, BGL2, PR5, SID2, EDS5, PAD4*) accumulated locally and systemically (*PR1, BGL2, PR5*) during SLWF nymph feeding. In contrast, jasmonic acid (JA)- and ethylene-dependent RNAs (*PDF1.2, VSP1, HEL, THI2.1, FAD3, ERS1, ERF1*) were repressed or not modulated in SLWF-infested leaves (Zarate et al. 2007).

A supporting evidence for the concept of plant vaccination comes in the form of the phenomenon of "priming" of plant defenses (Frost et al. 2008) that results in an enhanced and faster upregulation of genes that respond to the herbivore attack as compared to the plants that are non-primed. Plant priming of defenses can occur with direct herbivore attack as observed *in vitro* in case of *P. rapae* caterpillar attacks that were seen to not only prime Arabidopsis plants for induced resistance responses but also increase the spectrum of resistance against diverse pathogens that were otherwise ineffectively controlled (De Vos et al. 2006). Specifically, the caterpillar feeding was seen to stimulate the production of jasmonic acid (JA) and ethylene (ET) that results in significant reduction in disease incidence by bacterial pathogens, *Pseudomonas syringae* pv tomato and *Xanthomonas campestris* pv *armoraciae*. In addition, the grazing of leaf tissue by the herbivores primes the augmented expression of salicylic acid (SA)-dependent defenses and analysis revealed that the increased levels of ET act synergistically with SA-induced *PATHOGENESIS RELATED-1* (*PR-1*) to sensitize the tissues to develop resistance to the biotroph Turnip crinkle virus (TCV) that is SA-dependent defense response (De Vos et al. 2006). Exposing the plant tissues directly to volatile organic compounds (VOCs) was responsible for priming and inducing the defense mechanisms of the exposed "emitter" plant as well as the neighboring "receiver" plant as seen in studies conducted with lima bean plants treated with VOCs of beetle-damaged conspecific shoots (Heil and Bueno 2007) and in maize plants that expressed a specific subset of 10 JA-induced defence genes (Ton et al. 2007).

Ecological Paradigm of Plant-Insect Interactions

Induced resistance is a very effective tool for developing plant disease resistance against a wide spectrum of diseases, but the utilization of this technique entails high costs for allocating resources and the added limitation of the toxicity of the utilizable defensive compounds both for the plant as well as the environment. In this scenario, the process of plant priming proved to be an ecological, environmental friendly tool for enhancing natural plant disease resistance and an effective tool to transfer from lab to field level (van Hulten et al. 2006). Specifically, the process of plant defense priming not only equaled protection provided by insecticide treated plants (e.g., benzothiadiazole-treated wild type Arabidopsis compared to primed plants, van Hulten et al. 2006), but primed plants displayed significantly higher levels of fitness in the field than non-induced plants and plants expressing chemically or constitutively-induced direct defense (van Hulten et al. 2006). Further, plant defense priming, as clearly shown in the case of transcriptome analysis of about 8000 genes of rhizobacteria-induced systemic resistance (ISR) studies of Arabidopsis, showed the involvement of a limited gene expression and phytohormone induction, especially in tissues that have not been exposed to the priming agent, thus further strengthening the hypothesis of enhanced plant fitness conferred due to plant defense priming (Verhagen et al. 2004). In one of the best documented field study of interspecies signaling due to above-ground VOCs between native tobacco (*Nicotiana attenuata*) transplanted adjacent to clipped sagebrush (*Artemesia tridentata*), it was shown that Clipped sagebrush releases many biologically active VOCs, including methyl jasmonate (MeJA), methacrolein and a series of terpenoid and green leaf VOCs (Kessler et al. 2006). The study also involves observation of increases in fitness parameters and that priming (rather than direct elicitation) of native *N. attenuata*'s induced chemical defenses by a sagebrush-released VOCs (Kessler et al. 2006). An additional finding that showed the benefits of priming was that of an accelerated production of trypsin proteinase inhibitors when *Manduca sexta* caterpillars fed on plants previously exposed to clipped sagebrush resulting in lower total herbivore damage to plants exposed to clipped sagebrush and in a higher mortality rate of young Manduca caterpillars. This demonstrates priming of plant defense responses as a mechanism of plant-plant signaling in natural conditions, and more specifically, provides a unique and useful research opportunity for the analysis of interspecies plant signaling under ecologically realistic conditions (Kessler et al. 2006).

In ecological research, one of the important goals and focus is to understand how the phenotypic and genotypic traits of a plant species contribute to its fitness in terms of reproductive success. Molecular genetic approaches now enable ecological researchers opportunities involving sequencing of plant genomes and the availability of well-characterized mutants coupled with transgenic approaches in altering traits that mediate plant-insect interactions to address the ecological function of individual traits in a very precise fashion (Dicke et al. 2004; Kessler et al. 2004; Steppuhn et al. 2004). Research in molecular model species such as Arabidopsis now have valuable added information available in terms of induced defense through a molecular genetic approach (extensively reviewed in Van Poecke 2007). Arabidopsis, as a model species enabled researchers to dissect the understanding of gene functioning and also transferring certain useful genes such as terpene synthase genes from heterologous (non-brassicaceous) plants to Arabidopsis to investigate the ecological effects of the volatile terpenoid products of the terpene synthase in terms of attraction of carnivorous arthropods (Kappers et al. 2005; Schnee et al. 2006). It should be noted, however, that Arabidopsis is not the ideal plant for investigating the ecology of insect-plant interactions due to its early-season phenology and recent research involving other crucifer species such as *Cardamine cordifoilia* (Louda and Rodman 1996), *Lepidium virginicum* (Agrawal 2000), *Brassica oleracea, Brassica nigra* (van Dam et al. 2004) is providing more effective research opportunities investigate the effects of genes on community ecology. Research involving ecology of *B. oleracea* was for agricultural varieties (Shiojiri et al. 2002) and for feral and native populations (Moyes et al. 2000; Bukovinszky et al. 2008) and this provides unique research properties for an ecogenomic approach to community ecology of plant-insect interactions.

Other plant species such as those from Solanacea family such as *N. attenuata* and *Solanum nigrum* are being utilized for molecular ecological studies of community dynamics (Kessler et al. 2004; Schmidt et al. 2004). Specific studies in blocking the jasmonate cascade by silencing the LIPOXYGENASE3 gene in these species had important environmental consequences for the insect community associated with *N. attenuate* and some herbivores that were never before encountered on this plant were now able to colonize the plant (Kessler et al. 2004). Other gene silencing studies involving the CORONATINE INSENSITIVE1 gene enabled investigations in examining changes in a range of defense-related characteristics and changes in interactions with the natural community (Paschold et al. 2007).

Conclusion

The genetic, biochemical and molecular networks of plant-insect interactions are model platforms for integrative biological research especially in the emerging discipline of systems biology. The integration of the ecological paradigm further adds to the potential pioneer shifting research advances in molecular genetics and community ecology. Plant-insect interaction research studies extend the plant biology research beyond model species such as Arabidopsis into new model species such as those from the brassica species. Modern cutting edge research tools such as RNA interference, and virus-induced gene silencing enabled generation of specific lines where individual genes knockouts provided ecologists to integrate the molecular biology realm into plant disease resistance research (Paschold et al. 2007). This enabled the investigations of the effects of individual disease resistant genes on individual plant-insect interactions in terms of community dynamics.

References

Agrawal, A.A. and R. Karban. 1999. Why induced defenses may be favored over constitutive strategies in plants. pp. 45–61. *In*: Tollrian, R. and C.D. Harvell (eds.). The Ecology and Evolution of Inducible Defenses. Princeton University Press, Princeton.

Agrawal, A.A. 2000. Benefits and costs of induced plant defense for *Lepidium virginicum* (Brassicaceae). Ecology. 81: 1804–1813.

Arimura, G., C. Kost and W. Boland. 2005. Herbivore-induced, indirect plant defences. Biochim Biophys Acta. 1734: 91–111.

Bezemer, T.M. and N.M. van Dam. 2005. Linking aboveground and belowground interactions via induced plant defenses. Trends Ecol Evol. 20: 617–624.

Bodenhausen, N. and P. Reymond. 2007. Signaling pathways controlling induced resistance to insect herbivores in Arabidopsis. Mol Plant Microbe Interact. 20: 1406–1420.

Broekgaarden, C., E.H. Poelman, G. Steenhuis et al. 2007. Genotypic variation in genome-wide transcription profiles induced by insect feeding: *Brassica oleracea-Pieris rapae* interactions. BMC Genomics. 8: 239.

Bruinsma, M. and M. Dicke. 2008. Herbivore-induced indirect defence: from induction mechanisms to community ecology. pp. 31–60. *In*: Schaller, A. (ed.). Induced Plant Resistance to Herbivory. Springer Publishers, Berlin.

Bukovinszky, T., F.J.F. van Veen, Y. Jongema et al. 2008. Direct and indirect effects of resource quality on food web structure. Science. 319: 804–807.

Cardoza, Y.J., P.E.A. Teal and J.H. Tumlinson. 2003. Effect of peanut plant fungal infection on oviposition preference by *Spodoptera exigua* and on host searching behavior by *Cotesia marginiventris*. Environ Entomol. 32: 970–976.

D'Alessandro, M. and T.C.J. Turlings. 2006. Advances and challenges in the identification of volatiles that mediate interactions among plants and arthropods. Analyst. 131: 24–32.

De Vos, M., V.R. Van Oosten, R.M.P. Van Poecke et al. 2005. Signal signature and transcriptome changes of Arabidopsis during pathogen and insect attack. Mol Plant Microbe Interact. 18: 923–937.

De Vos, M., W. Van Zaanen, A. Koornneef et al. 2006. Herbivore-induced resistance against microbial pathogens in Arabidopsis. Plant Physiol. 142: 352–363.

Dick, M., R.M.P. van Poecke and J.G. de Boer. 2003. Inducible indirect defence of plants: from mechanisms to ecological functions. Basic Appl Ecol. 4: 27–42.

Dicke, M. and R.M.P. Van Poecke. 2002. Signalling in plant-insect interactions: signal transduction in direct and indirect plant defence. pp. 289–316. *In*: Scheel, D. and C. Wasternack (eds.). Plant Signal Transduction. Oxford University Press, Oxford.

Dicke, M., J.J.A. van Loon and P.W. de Jong. 2004. Ecogenomics benefits community ecology. Science. 305: 618–619.

Ehrlich, P.R. and P.H. Raven. 1964. Butterflies and plants: a study in coevolution. Evolution Int J Org Evolution. 18: 586–608.

Frost, C.J., M.C. Mescher, J.E. Carlson et al. 2008. Plant defense priming against herbivores: getting ready for a different battle. Plant Physiol. 146: 818–824.

Gols, R., T. Bukovinszky, L. Hemerik et al. 2005. Reduced foraging efficiency of a parasitoid under habitat complexity: implications for population stability and species coexistence. J Anim Ecol. 74: 1059–1068.

Harvey, J.A., N.M. vanDam and R. Gols. 2003. Interactions over four trophic levels: food plant quality affects development of a hyperparasitoid as mediated through a herbivore and its primary parasitoid. J Anim Ecol. 72: 520–531.

Heidel, A.J. and I.T. Baldwin. 2004. Microarray analysis of salicylic acid and jasmonic acid-signaling in responses of *Nicotiana attenuata* to attack by insects from multiple feeding guilds. Plant Cell Environ. 27: 1362–1373.

Heil, M., S. Greiner, H. Meimberg et al. 2004. Evolutionary change from induced to constitutive expression of an indirect plant resistance. Nature. 430: 205–208.

Heil, M. and J.C.S. Bueno. 2007. Within-plant signaling by volatiles leads to induction and priming of an indirect plant defense in nature. Proc Natl Acad Sci USA. 104: 5467–5472.

Howe, G.A. 2004. Jasmonates as signals in the wound response. J Plant Growth Regul. 23: 223–237.

Kahl, J., D.H. Siemens, R.J. Aerts et al. 2000. Herbivore-induced ethylene suppresses a direct defense but not a putative indirect defense against an adapted herbivore. Planta. 210: 336–342.

Kaplan, I. and R.F. Denno. 2007. Interspecific interactions in phytophagous insects revisited: a quantitative assessment of competition theory. Ecol Lett. 10: 977–994.

Kappers, I.F., A. Aharoni, T. van Herpen et al. 2005. Genetic engineering of terpenoid metabolism attracts, bodyguards to Arabidopsis. Science. 309: 2070–2072.

Kempema, L.A., X.P. Cui, F.M. Holzer et al. 2007. Arabidopsis transcriptome changes in response to phloem-feeding silverleaf whitefly nymphs. Similarities and distinctions in responses to aphids. Plant Physiol. 143: 849–865.

Kessler, A. and I.T. Baldwin. 2002. Plant responses to insect herbivory: the emerging molecular analysis. Annu Rev Plant Biol. 53: 299–328.

Kessler, A. and I.T. Baldwin. 2004. Herbivore-induced plant vaccination. Part I. The orchestration of plant defenses in nature and their fitness consequences in the wild tobacco *Nicotiana attenuata*. Plant J. 38: 639–649.

Kessler, A., R. Halitschke and I.T. Baldwin. 2004. Silencing the jasmonate cascade: induced plant defenses and insect populations. Science. 305: 665–668.

Kessler, A., R. Halitschke, C. Diezel et al. 2006. Priming of plant defense responses in nature by airborne signaling between *Artemisia tridentata* and *Nicotiana attenuata*. Oecologia. 148: 280–292.

Kessler, A. and R. Halitschke. 2007. Specificity and complexity: the impact of herbivore-induced plant responses on arthropod community structure. Curr Opin Plant Biol. 10: 409–414.

Koornneef, A. and C.M.J. Pieterse. 2008. Cross talk in defense signaling. Plant Physiol. 146: 839–844.

Krischik, V.A., P. Barbosa and C.F. Reichelderfer. 1988. Three trophic level interactions: allelochemicals, *Manduca sexta* (L.), and *Bacillus thuringiensis* var. kurstaki Berliner. Environ Entomol. 17: 476–482.

Kusnierczyk, A., P. Winge, T.S. Jorstad et al. 2008. Towards global understanding of plant defence against aphids—timing and dynamics of early Arabidopsis defence responses to cabbage aphid attack. Plant Cell Environ. 31: 1097–1115.

Li, C.Y., M.M. Williams, Y.T. Loh et al. 2002. Resistance of cultivated tomato to cell content-feeding herbivores is regulated by the octadecanoid-signaling pathway. Plant Physiol. 130: 494–503.

Louda, S.M. and J.E. Rodman. 1996. Insect herbivory as a major factor in the shade distribution of a native crucifer (*Cardamine cordifolia* A. Gray, bittercress). J Ecol. 84: 229–237.

Mercke, P., I.F. Kappers, F.W.A. Verstappen et al. 2004. Combined transcript and metabolite analysis reveals genes involved in spider mite induced volatile formation in cucumber plants. Plant Physiol. 135: 2012–2024.

Moyes, C.L., H.A. Collin, G. Britton et al. 2000. Glucosinolates and differential herbivory in wild populations of Brassica oleracea. J Chem Ecol. 26: 2625–2641.

Mumm, R. and M. Hilker. 2006. Direct and indirect chemical defence of pine against folivorous insects. Trends Plant Sci. 11: 351–358.

Paschold, A., R. Halitschke and I.T. Baldwin. 2007. Co(i)-ordinating defenses: NaCOI1 mediates herbivore-induced resistance in *Nicotiana attenuate* and reveals the role of herbivore movement in avoiding defenses. Plant J. 51: 79–91.

Pieterse, C.M.J., S.C.M. Van Wees, J. Ton et al. 2002. Signaling in rhizobacteria-induced systemic resistance in *Arabidopsis thaliana*. Plant Biol. 4: 535–544.

Reymond, P. and E.E. Farmer. 1998. Jasmonate and salicylate as global signals for defense gene expression. Curr Opin Plant Biol. 1: 404–411.

Reymond, P., H. Weber, M. Damond et al. 2000. Differential gene expression in response to mechanical wounding and insect feeding in *Arabidopsis*. Plant Cell. 12: 707–719.

Reymond, P., N. Bodenhausen, R.M.P. Van Poecke et al. 2004. A conserved transcript pattern in response to a specialist and a generalist herbivore. Plant Cell. 16: 3132–3147.

Rodriguez-Saona, C., J.A. Chalmers, S. Raj et al. 2005. Induced plant responses to multiple damagers: differential effects on an herbivore and its parasitoid. Oecologia. 143: 566–577.

Schmidt, D.D., A. Kessler, D. Kessler et al. 2004. *Solanum nigrum*: a model ecological expression system and its tools. Mol Ecol. 13: 981–995.

Schnee, C., T.G. Kollner, M. Held et al. 2006. The products of a single maize sesquiterpene synthase form a volatile defense signal that attracts natural enemies of maize herbivores. Proc Natl Acad Sci USA. 103: 1129–1134.

Schoonhoven, L.M., J.J.A. van Loon and M. Dicke. 2005. Insect-Plant Biology, Ed 2. Oxford University Press, Oxford.

Shiojiri, K., J. Takabayashi, S. Yano et al. 2002. Oviposition preferences of herbivores are affected by tritrophic interaction webs. Ecol Lett. 5: 186–192.

Snoeren, T.A.L., P.W. De Jong and M. Dicke. 2007. Ecogenomic approach to the role of herbivore-induced plant volatiles in community ecology. J Ecol. 95: 17–26.

Steppuhn, A., K. Gase, B. Krock et al. 2004. Nicotine's defensive function in nature. PLoS Biol. 2: e217.

Stork, N.E. 2007. Biodiversity: world of insects. Nature. 448: 657–658.

Thaler, J.S. 1999. Jasmonate-inducible plant defenses cause increased parasitism of herbivores. Nature. 399: 686–688.

Thines, B., L. Katsir, M. Melotto et al. 2007. JAZ repressor proteins are targets of the SCFCO11 complex during jasmonate signalling. Nature. 448: 661–665.

Ton, J., M. D'Alessandro, V. Jourdie et al. 2007. Priming by airborne signals boosts direct and indirect resistance in maize. Plant J. 49: 16–26.

van Dam, N.M., L. Witjes and A. Svatos. 2004. Interactions between aboveground and belowground induction of glucosinolates in two wild Brassica species. New Phytol. 161: 801–810.

van Hulten, M., M. Pelser, L.C. van Loon et al. 2006. Costs and benefits of priming for defense in Arabidopsis. Proc Natl Acad Sci USA. 103: 5602–5607.

Van Poecke, R.M.P. 2007. Arabidopsis-insect interactions. *In*: Somerville, C.R. and E.M. Meyerowitz (eds.). The Arabidopsis Book. American Society of Plant Biologists, Rockville, MD, doi/10.1199/tab.0107, www.aspb.org/publications/arabidopsis/.

Verhagen, B.W.M., J. Glazebrook, T. Zhu et al. 2004. The transcriptome of rhizobacteria-induced systemic resistance in Arabidopsis. Mol Plant Microbe Interact. 17: 895–908.

Voelckel, C. and I.T. Baldwin. 2004a. Generalist and specialist lepidopteran larvae elicit different transcriptional responses in *Nicotiana attenuata*, which correlate with larval FAC profiles. Ecol Lett. 7: 770–775.

Voelckel, C. and I.T. Baldwin. 2004b. Herbivore-induced plant vaccination. Part II. Array-studies reveal the transience of herbivore-specific transcriptional imprints and a distinct imprint from stress combinations. Plant J. 38: 650–663.

Walling, L.L. 2000. The myriad plant responses to herbivores. J Plant Growth Regul. 19: 195–216.

Will, T., W.F. Tjallingii, A. Thonnessen et al. 2007. Molecular sabotage of plant defense by aphid saliva. Proc Natl Acad Sci USA. 104: 10536–10541.

Zarate, S.I., L.A. Kempema and L.L. Walling. 2007. Silverleaf whitefly induces salicylic acid defenses and suppresses effectual jasmonic acid defenses. Plant Physiol. 143: 866–875.

Zheng, S.J., J.P. van Dijk, M. Bruinsma et al. 2007. Sensitivity and speed of induced defense of cabbage (*Brassica oleracea* L.): dynamics of BoLOX expression patterns during insect and pathogen attack. Mol Plant Microbe Interact. 20: 1332–1345.

CHAPTER 3

The Cotton-Insect Interactive Transcriptome

Molecular Elements Involved in Plant-Insect Interactions

Mirzakamol S. Ayubov and *Ibrokhim Y. Abdurakhmonov**

INTRODUCTION

Insects are the most diverse and adapted species of animals on our planet. Insects can live in all habitats of the earth, from the ocean floor to the peak of a mountain (Imms 1964) and play a significant role in plant and animal life. Insects have been classified as predators, anti-biological insect-pests, pollinators or producers of valuable products (Offor et al. 2014). In addition, they are used for pharmacological purposes such as anti-venoms or antibodies (Moreno and Giralt 2015). Among all known insects, only fifty percent of them are pests, but a significant few pose a serious threat to agriculture and food security (Offor et al. 2014).

Cotton (*Gossypium* sp.) is susceptible to a wide range of insect pests. There are different types of pests that are harmful to cotton, such as cotton bollworm (*Helicoverpa armigera*), plant bugs (*Miridae* sp.), stink bugs (*Halyomorpha halys*), aphids (*Aphis gossypii* Glover), thrips (*Thysanoptera* sp.) and spider mites (*Tetranychidae*). Cotton bollworm, specifically, the pink bollworm (*Pectinophora gossypiella*) brings more damage to crop production by damaging the bolls (Xiong et al. 2015; Tassone et al. 2016), and *Lygus lineolaris* (*Palisot de Beauvois*) is a polyphagous, phytophagous insect pest in cotton and other important crops (Showmaker 2016).

Center of Genomics and Bioinformatics, Academy of Sciences the Republic of Uzbekistan, University Street-2, Qibray region, Tashkent District, 111215 Uzbekistan.
Email: mirzakamol.ayubov@genomics.uz
* Corresponding author: genomics@uzsci.net; ibrokhim.abdurakhmonov@genomics.uz

The chewing insect boll weevil larvae (*Anthonomus grandis*) and phloem feeding insect *Bemisia tabaci* (whitefly) (Artico et al. 2014; Li et al. 2016) cause the loss of diverse cotton products.

Although researchers have elucidated many aspects of insect-host interactions and developed effective tools against the insect pests including genetic engineering of *Bacillus thuringiensis* (or Bt) Cry toxin gene and using RNA interference of vitally important insect genes (Abdurakhmonov et al. 2016), many aspects of molecular mechanisms of insect resistance in cotton require a more in-depth study with application of new generation of "omics" tools such as transcriptome analysis. In past years, to better understand molecular mechanisms in cotton plant-insect interactions, the transcriptome of insect venom and midgut extracts as well as olfactory systems were sequenced together with plant damaged tissues. Recent research revealed that most transcriptome related research studies have been dedicated to the analysis of cotton bollworm *Helicoverpa armigera*. This insect is one of the widely-spread pests feeding on hundreds of varieties of plant species (Liu et al. 2014), and causes large-scale damage to agricultural production. Various species of bollworm cause some of the greatest yield losses in Africa annually even in Bt-cotton plantations (Rangarirai et al. 2015; Liu et al. 2014). Hence, transcriptome analysis is a research priority directed not only to solve the problems resulting by a pest such as *Helicoverpa armigera* but also to understand major insect resistance pathways in a cash crop such as cotton.

Transcriptome Analysis of Insect-Insecticide Interactions

In conventional farming, most of the cotton pests were controlled by using insecticides. Insecticides differ by their targets, for example, ovicides target insect eggs, while larvicides target larvae. However, during the last decade, the usage of insecticides did not prove to be an efficient strategy. Efficiency of insecticides over time can be evaluated by resistance developed by insects. Even though the expression Cry1Ac gene from *Bacillus thuringiensis* in transgenic cotton cultivars led to significantly lower usage of pesticides for many years (Qiu et al. 2015; Cao et al. 2016), in subsequent years, several pests including green mirid bugs (GMB), *Apolygus lucorum* and *Adelphocoris suturalis* Jakovlev, have increased their resistance to Bt toxin. Therefore, *de novo* transcriptome assembly and gene expression analyses have been implemented to understand the molecular, genetic, biochemical and physiological mechanisms of toxin resistance of these insects during different developmental stages (Cao et al. 2016; Tian et al. 2015). To identify target genes, the transcriptome of all developmental stages of the insect pest *A. grandis* was analyzed and several key insect genes (e.g., chitin synthase 1) have been characterized (Firmino et al. 2013).

The pink cotton bollworm *Pectinophora gossypiella* is another model organism for insect responses to Bt toxins, but the molecular mechanism of its tolerance was studied to a lesser extent. Using *de novo* transcriptome assembly for the midgut of *P. gossypiella*, 46,458 transcripts have been derived from 39,874 unigenes ("Unigene" is the NCBI transcriptome database). The transcriptome data presented those relevant midgut proteins critical for xenobiotic detoxification, nutrient digestion and allocation, as well as for the discovery of protein receptors important for Bt intoxication (Tassone et al. 2016).

Interactions between secondary metabolites of cotton and *H. armigera* insect resistance was unclear before Tao (2012) who studied enzyme cytochrome P450 monooxygenases (P450s) in cotton bollworm that increases detoxification of gossypol (Tao et al. 2012). Subsequently, comparative detoxification enzymes of *H. assulta* and *H. armigera*, as well as transcriptome of *Aphis gossypii* Glover were investigated (Li et al. 2013a,b). Five gossypol-induced P450s genes contributed to cotton bollworm tolerance to deltamethrin insecticide. When one of the genes CYP9A14 was knocked down by plant-mediated RNA interference (RNAi), the larvae exhibited more sensitivity to the insecticide (Tao et al. 2012). For increasing adaptation to very different feeding sources, *A. grandis* was able to produce transcripts encoding proteins involved in catalytic processes of macromolecules. Besides, these proteins are involved in detoxification mechanisms such as expression of p450 genes, glutathione-S-transferase, and carboxylesterases (Salvador et al. 2014).

Effects of systemic insecticides, thiamethoxam and spirotetramat, belonging to the class of neonicotinoids widely used against *A. gossypii* Glover were also investigated using Illumina-Solexa sequencing technology (Bentley et al. 2008). A total of 22,569,311 and 21,317,732 clean reads (sequences "cleaned" of vector and adaptor sequences) were obtained from the thiamethoxam-resistant strain (ThR) and susceptible strain (SS) transcriptomes, as well as total of 22,430,522 and 21,317,732 clean reads from RS (resistant strain) and SS cDNA libraries. Studies revealed that unigenes significantly changed in both ThR and RS libraries compared to the both SS strains (Pan et al. 2015a,b). Transcriptomic sequences were used to analyze the involvement of potential heat shock protein (Hsp) homologs of *Latrodectus hesperus* (the western black widow spider or western widow) proteins in thermal stress response transcriptome analysis. Results showed the up regulation of Hsp70, Hsp40, and two small Hsps in the heat-challenged adults of *L. hesperus*. These results helped in understanding the fundamental mechanisms of abiotic stress response genes and their role in thermotolerance (Hull et al. 2013).

Heat challenged expression levels of ribosomal proteins, heat shock protein 70 (Hsp70), ATP synthase, ecdysteroid UDP-glucosyltransferase and esterase in aphids were up-regulated significantly in the ThR and SR strains compared to the SS strains. The decreased expression of genes encoding cuticle proteins, salivary glue proteins, and energy ATP synthase, and cytochrome c oxidase, fibroin heavy chain was observed (Pan et al. 2015a,b). A nicotinic acetylcholine receptor *(nAChR)* α subunit was down-regulated in the ThR strain (Pan et al. 2015a). Among the differentially expressed genes (DEGs) for cytochrome P450, *6A2* was the only up-regulated gene in the SR strain (Pan et al. 2015b). These data illustrated that genetic changes in *nAChR* genes and up-regulated ribosomal proteins, ecdysteroid UDP-glucosyltransferase, cytochrome c oxidase, esterase and peroxidase may confer the resistance of cotton aphids to thiamethoxam (Pan et al. 2015a). Subsequently, an over-expression of *CYP6A2* gene associated with spirotetramat resistance and cross-resistance in the resistant strain of *Aphis gossypii* Glover was achieved. Suppression of CYP6A2 transcripts by RNAi significantly increased the sensitivity of the resistant aphid to spirotetramat (Peng et al. 2016).

The increased quantity of secondary metabolites, namely, natural insecticides in plants is associated with abiotic stresses such as treatment with NaCl. The increase in levels of gossypol, flavonoids and tannic acid in plants was linked to reduced aphid population when cotton plant was treated with 50–200 mM NaCl. Compared to non-treated samples, a salt-treated aphid transcriptome analysis showed higher expressions of genes including fatty acid and lipid biosynthesis, carbohydrate and amino acid metabolism, energy metabolism and few others responsive for cell motility pathway (Wang et al. 2015). qRT-PCR showed a high expression of transcripts for *CYP6A14, CYP6A13, CYP303A1, NADH* and fatty acid synthase genes. In contrast, *CYP307A1* and two ecdysone-induced protein genes were down regulated. The study also showed expression of genes related to growth and development of aphids being highly expressed by enhanced secondary metabolism in cotton under salt stress. The involvement of aphid gene *CYP307A1* in ecdysone synthesis showed its positive correlation with the population dynamics (Wang et al. 2015).

Transcriptome Analysis of Olfactory Genes in Diverse Insect Species

Olfaction is a significantly essential factor in the life cycle of insects. Insects sense the smell through their antennae, the transcriptome of which can serve as a good model to understand the olfaction mechanisms in insects.

In *Adelphocoris suturalis*, a plant bug that is one of the most serious insect pests of Bt cotton in China, two soluble protein compositions, namely, odorant binding proteins (OBPs) and chemosensory proteins (CSPs) participate in the initial biochemical recognition steps in semi-chemical perception and insect olfactory signal transduction (Gu et al. 2013; Cui et al. 2016). Transcriptome of three asexual developmental stages (wingless spring and summer morphs and winged adults) of *Aphis gossypii* Glover have been compared and characterized. The number and length of introns observed were much higher in general and this appears to be a unique feature of Aphid OBP and CSP genes. On the other hand, higher abundance of CSP transcripts than OBP is another unique feature in aphids (Gu et al. 2013). Different transcripts were expressed in male and female species. For instance, 4 OBPs (AsutOBP1, 4, 5 and 9) and 1 CSP (AsutCSP1) were expressed at higher levels in male than in female antennae in *A. suturalis* (Cui et al. 2016).

Many candidate chemosensory genes were identified by sequencing female- and male-antennae transcriptomes of *Chrysopa pallens*, *Spodoptera littoralis* and *Chrysoperla sinica* including OBPs, CSPs, odorant receptors (ORs), and ionotropic receptors (IRs) (Li et al. 2013c; Poivet et al. 2013; Li et al. 2015b). Existence of three types of chemosensory receptors including ORs, gustatory receptors (GRs) and IRs in insects were identified by investigating expansion of a bitter taste receptor family in a polyphagous insect herbivore.

One hundred and ninety-seven novel GR genes were also identified from the polyphagous pest *Helicoverpa armigera* (Liu et al. 2014; Xu et al. 2016). These receptors play vital roles in sensing chemical signals, sex-specific or developmental stage-specific chemosensory behaviors that guide insect behaviors. Using transcriptome sequencing, trinity RNA-seq assemblies and extensive manual curation, 60 candidate ORs, 10 GRs and 21 IRs have been identified in *H. armigera* (Liu et al. 2014). Further, 83 and 68 transcripts related to olfaction have been identified *H. armigera* and *H. assulta*. Moreover, more than a thousand transcripts of digestive enzymes were identified in the same insect species. Comparative analysis showed that detoxification enzymes, e.g., P450, carboxypeptidase, and ATPase were higher in *H. assulta* than in *H. armigera*. These detoxification enzymes would help them to increase the food detoxification and utilization efficiency (Li et al. 2013b).

As in other organisms, innate immunity is important for defense of *H. armigera* from invading pathogens. Both fat body and hemocytes are important organs involved in the immune response (Xiong et al. 2015). *De novo* sequencing and transcriptome analysis of endoparasitoid

Aenasius arizonensis showed the involvement of venom glands in host developmental arrest, disrupting the host immune system, and host paralysis (Shaina et al. 2016). After pathogen infections, it was observed that the immunotranscriptome of *H. armigera* larvae and the related gene expression have changed. Some immunity-related genes were activated in fungus and bacterium-challenged fat bodies, while others were suppressed in the fungus, *Beauveria bassiana,* challenged by hemocytes. More immunity-related genes were induced by the fat bodies than those by hemocytes (Xiong et al. 2015). To observe the involvement of gene families in immunity between fat bodies and antennae, Legeai's study (2014) was very important in *Spodoptera frugiperda*. Sf_TR2012b transcriptome enabled researchers to explore the spatial and temporal expression of genes and to observe that some olfactory receptors are expressed in antennae and palps but also in other non-related tissues such as fat bodies (Legeai et al. 2014). The naturally occurring ecdysteroid hormone which controls the ecdysis (moulting) and metamorphosis of insects, 20-hydroxyecdysone (20E), influences innate immunity and induced the expression of the immunity-related genes in many holometabolous insects as well as in the cotton bollworm. This has been proved by transcriptome analysis of peptidoglycan-challenged fat body of cotton bollworm. Results showed that antibacterial activities were enhanced, mRNA levels of pattern recognition receptors and antimicrobial peptides significantly increased during the wandering or pupal stage of the insect life cycle (Wang et al. 2014).

Transcript Involvement in Pheromone Biosynthesis

Pheromones are produced in metathoracic scent glands (MTGs) in *A. suturalis* Jakovlev and play an important role in survival and population propagation of the insect. There was very little information about the molecular basis of the pheromone biosynthesis of the insects till recently (Luo et al. 2014). It is essential to clarify the involvement of genes in the production of pheromone components. Two similar sex pheromone components Z9-16:Ald and Z11-16:Ald have been found in both the sibling species cotton bollworm (*H. armigera*) and oriental tobacco budworms (*H. assulta*). Differences in sex pheromone component can lead to reproductive isolation and it is genetically controlled. To investigate how the ratios of the pheromone components are differently regulated in the two species, cDNA libraries have been sequenced from the pheromone glands of *H. armigera* and *H. assulta*. The research highlighted the involvement of some of the transcripts among the identified ones in the sex pheromone biosynthesis pathways (Li et al. 2015a). qRT-PCR results demonstrated higher expression levels of *Asdelta9-DES*, *AsFAR*, *AsAOX*, *Ascarboxylesterase*, *AsNT-ES* and *AsATFs* genes in the releasing period of sex pheromones in female *A. suturalis* Jakovlev. Thus, many potential pheromone biosynthetic pathway genes were identified (Luo et al. 2014).

Transcriptome Analysis Under Herbivore Stresses

Insect herbivores destroy world's major crops and thus affect global economic growth. High yielding crop varieties have been developed for the past three decades by novel breeding techniques (Kerin 1994). In the process of artificial selection, suitable crops are largely selected by breeders for increased agricultural productivity and human consumption. In examining the genetic bottlenecks of the process, an important parameter continues to be the ability of every plant species to have its own mechanism to protect itself from herbivore attacks. The involvement of many genes in the biotic stress responses such as a mitogen-activated protein kinase (MAPK), transcription factors (WRKY and ERF protein domains) and signaling by ethylene (ET) and jasmonic acid (JA) hormones have been identified in *Anthonomus grandis* using transcriptome, differentially-expressed gene (DEG) sequence analysis and virus-induced gene silencing (VIGS) (Artico et al. 2014; Li et al. 2016). MAPK-WRKY-JA functions and ET pathways were suppressed by virus-induced gene silencing (VIGS) in *Bemisia tabaci* (Li et al. 2016). This results in the release of elevated levels of volatiles which can serve as a chemical signal. In response to the insect herbivore, plants directly and indirectly produce volatile organic compounds. However, the molecular basis for defense response during insect herbivory trigger in cotton plant and how the defensive compounds are manipulated as part of biological processes have not been well studied. Huang et al. (2015) showed the transcriptome changes and volatile characteristics of cotton plants in response to cotton bollworm (CBW). Around two thousand transcripts showed different expressions after CBW infestation. Cluster analysis indicated that CBW-induced genes play important roles in CBW-induced defenses (Huang et al. 2015).

Acyclic terpenes and the shikimate pathway product indole are biosynthesized *de novo* following insect damage and experiments demonstrated that the application of caterpillar oral secretions increased the production and release several volatiles that are synthesized *de novo* in response to insect feeding. The role of plants in mediating the interaction between herbivores and natural enemies of herbivores was demonstrated to be a critically dynamic phenomenon (Pare et al. 1997).

To better control the insects, a subunit of mitochondrial complex I NDUFV2 that catalyzes NADH dehydrogenation in respiratory chain was suppressed by double-stranded RNA (dsRNA). When cotton bollworm larvae were fed with transgenic cotton tissue, expression of NDUFV2 dsRNA caused mortality up to 80%, and no larvae survived. Comparative transcriptome analysis showed a repression of Dopa decarboxylase genes (Wu et al. 2016). At the same time, the role of insect predators that feed on insect-pests was examined. One of the insect predators *Arma chinensis,* effectively used to control several insect-pests including Colorado potato beetle, cotton bollworm, and mirid bugs, was fed with artificial diet (pig liver and tuna) to better understand the impact of such diets. Transcriptome analyses demonstrated the differential expressions of thousands of genes between pupae-fed and diet-fed *A. chinensis* which can be efficiently used for the reduction of the expected damage by insect-pests (Zou et al. 2013). These novel transcriptome approaches as vital applications in insect research should be very helpful to shed light on details of the molecular signatures and their interactions during insect-plant interaction.

Conclusions

This brief overview of research studies of the past decade (Table 1) highlighted herein reveals that the new generation "omics" approach such as transcriptome analysis has helped to better understand the molecular mechanisms of insect resistance in a globally important cash crop such as cotton. In particular, the transcriptome analyses for insecticide treatment, olfaction, and pheromone production processes in insects as well as research involving herbivore stresses in cotton has revealed novel information on key genes and molecular signatures that are essential to understand resistance mechanisms to help develop novel insect resistant cotton cultivars.

Acknowledgments

We are thankful to the Academy of Sciences of Uzbekistan and Committee for Coordination of Science and Technology Development for continuous grant funding of projects No. FA-F5-T030/FA-F-5-021; FA-A6-T085; I-2015-6-15/2 and I5- FQ-0-89-870, the Office of International Research Programs (OIRP) of the United States Department of Agriculture (USDA)—Agricultural Research Service (ARS), Texas A&M University and U.S. Civilian Research & Development Foundation (CRDF) for financial support of scientific projects No. P120, P120A, P121, P121B, UZB-TA-31016, UZB-TA-31017, and UZB-TA-2992 of cotton in Uzbekistan.

Table 1. Recent studied cotton-insect transcriptome analyses.

No	Targeted genes and proteins	Used methods	Insect name	References
1	*Asdelta9-DES*, *AsFAR*, *AsAOX*, *Ascarboxylesterase*, *AsNT-ES* and *AsATFs*	qRT-PCR	*Adelphocoris suturalis* Jakovlev	Luo (2014)
2	Development pathways, hormone biosynthesis, sex differences and wing formation	Quantitative real-time PCR	*Adelphocoris suturalis* Jakovlev	Tian (2015)
3	Odorant binding proteins (OBPs) and chemosensory proteins (CSPs)	Transcriptome	*Adelphocoris suturalis* Jakovlev	Cui (2016)
4	Calreticulin, Serine Protease Precursor and Arginine kinase proteins	*De novo* sequencing and transcriptome analysis	*Aenasius arizonensis*	Shaina (2016)
5	Mitogen-activated protein kinase (MAPK), transcription factors (WRKY and ERF) and signaling by ethylene (ET) and jasmonic acid (JA) hormones	Transcriptome, DEG sequence	*Anthonomus grandis*	Artico (2014)
6	Chitin synthase 1	RNA interference	*Anthonomus grandis*	Firmino (2013)
7	P450	Pyrosequencing	*Anthonomus grandis*	Salvador (2014)
8	Odorant binding proteins (OBPs) and chemosensory proteins	RNA-seq analyses	*Aphis gossypii* Glover	Gu (2013)
9	*nAChR*	Transcriptome	*Aphis gossypii* Glover	Pan (2015a)
10	Cuticle proteins, salivary glue protein, fibroin heavy chain, energy ATP synthase, and cytochrome c oxidase	Transcriptome	*Aphis gossypii* Glover	Pan (2015b)
11	*CYP6A14*, *CYP6A13*, *CYP303A1*, CYP307A1, *NADH* dehydrogenase and fatty acid synthase	qRT-PCR	*Aphis gossypii* Glover	Wang (2015)
12	CYP6A2	RNA interference	*Aphis gossypii* Glover	Peng (2016)
13	Cry1Ac	Overexpression	*Apolygus lucorum*	Cao (2016)
14	Dopa decarboxylase	dsRNA	*Apolygus lucorum*	Wu (2016)
15	Thousands of genes	Transcriptome	*Arma chinensis*	Zou (2013)

16	Mitogen-activated protein kinase (MAPK), transcription factors (WRKY and ERF) and signaling by ethylene (ET) and jasmonic acid (JA) hormones	Virus-induced gene silencing (VIGS)	*Bemisia tabaci (whitefly)*	Li(2016)
17	Chemosensory	Transcriptome	*Chrysopa pallens*	Li (2013c)
18	Chemosensory	Transcriptome	*Chrysoperla sinica*	Li (2015b)
19	CYP9A14	RNA interference	*Helicoverpa armigera*	Tao 2012
20	Detoxification enzyme, e.g., P450, carboxypeptidase, and ATPase	Transcriptome	*Helicoverpa armigera*	Li (2013b)
21	Gustatory receptor	RNA-seq analyses	*Helicoverpa armigera*	Liu (2014)
22	20-hydroxyecdysone (20E)	Transcriptome	*Helicoverpa armigera*	Wang (2014)
23	Cry1Ac	Overexpression	*Helicoverpa armigera*	Qiu (2015)
24	PGRP-SA1, Serpin1, Toll-14, and Spz2	RNA-seq analyses	*Helicoverpa armigera*	Xiong (2015)
25	Jasmonic acid	Cluster analysis	*Helicoverpa armigera*	Huang (2015)
26	HarmGR35, HarmGR50 and HarmGR195	Feeding	*Helicoverpa armigera*	Xu (2016)
27	Detoxification enzymes	High-throughput sequencing	*Helicoverpa armigera* and *Helicoverpa assulta*	Li (2013a)
28	Pheromone components Z9-16:Ald and Z11-16:Ald	Semi-quantitative RT-PCR, qRT-PCR	*Helicoverpa armigera* and *Helicoverpa assulta*	Li(2015a)
29	Heat shock protein (Hsp) homologs	Pyrosequencing, *de novo* assamble	*Lygus hesperus*	Hull (2013)
30	Salivary gland proteins	Sialotranscriptome	*Lygus lineolaris*	Showmaker (2016)
31	BT-toxin	*de novo* transcriptome	*Pectinophora gossypiella*	Tassone (2016)
32	The acyclic terpenes (E,E)-[alpha]-farnesene, (E)-[beta]-farnesene, (E)-[beta]-ocimene, linalool, (E)-4,8-dimethyl-1,3,7-nonatriene, and (E/E)-4,8,12-trimethyl-1,3,7,11-tridecatetraene	Biosynthesized *de novo*	*Spodoptera exigua Hubner*	Pare (1997)
33	Sf_TR2012b	Reference transcriptome	*Spodoptera frugiperda*	Legeai (2014)
34	Chemosensory	Transcriptome	*Spodoptera littoralis*	Poivet (2013)

References

Abdurakhmonov, I.Y., M.S. Ayubov, K.A. Ubaydullaeva et al. 2016. RNA interference for functional genomics and improvement of cotton (*Gossypium* spp.). Front Plant Sci. 7: 1–17.

Artico, S., M. Ribeiro-Alves, O.B. Oliveira-Neto et al. 2014. Transcriptome analysis of *Gossypium hirsutum* flower buds infested by cotton boll weevil (*Anthonomus grandis*) larvae. BMC Genomics. 15: 854.

Bentley, D.R., S. Balasubramanian, S.P. Swerdlow et al. 2008. Accurate whole human genome sequencing using reversible terminator chemistry. Nature. 456: 53–59.

Cao, D.P., Y. Liu, J.J. Wei et al. 2016. A *de novo* transcriptomic analysis to reveal functional genes in *Apolygus lucorum*. Insect Sci. 23: 2–14.

Cui, H.H., S.H. Gu, X.Q. Zhu et al. 2016. Odorant-binding and chemosensory proteins identified in the antennal transcriptome of *Adelphocoris suturalis* Jakovlev. Comp Biochem Physiol Part D Genomics Proteomics. pii: S1744-117X(16)30018-1.

Dubey, N.K., R. Goel, A. Ranjan et al. 2013. Comparative transcriptome analysis of *Gossypium hirsutum* L. in response to sap sucking insects: aphid and whitefly. BMC Genomics. 14: 241.

Firmino, A.A., F.C. Fonseca, L.L. de Macedo et al. 2013. Transcriptome analysis in cotton boll weevil (*Anthonomus grandis*) and RNA interference in insect pests. PLoS One. 8: e85079.

Gu, S.H., K.M. Wu, Y.Y. Guo et al. 2013. Identification and expression profiling of odorant binding proteins and chemosensory proteins between two wingless morphs and a winged morph of the cotton aphid *Aphis gossypii* glover. PLoS One. 8: e73524.

Huang, X.Z., J.Y. Chen, H.J. Xiao et al. 2015. Dynamic transcriptome analysis and volatile profiling of *Gossypium hirsutum* in response to the cotton bollworm *Helicoverpa armigera*. Sci Rep. 5: 11867.

Hull, J.J., S.M. Geib and J.A. Fabrick. 2013. Sequencing and *de novo* assembly of the western tarnished plant bug (*Lygus hesperus*) transcriptome. PLoS One. 8: e55105.

Imms, A.D. 1964. Outlines of Entomology. 5th ed. pp. 224. Methuen. London, UK.

Kerin, J. 1994. Opening address. Proceedings of the 6th International Working Conference on Stored-product Protection, 17–23 April 1994. Volume 1. pp. xix–xx. Canberra, Australia.

Legeai, F., S. Gimenez, B. Duvic et al. 2014. Establishment and analysis of a reference transcriptome for Spodoptera frugiperda. BMC Genomics. 15: 704.

Li, H., H. Zhang, R. Guan et al. 2013. Identification of differential expression genes associated with host selection and adaptation between two sibling insect species by transcriptional profile analysis. BMC Genomics. 14: 582.

Li, J., L. Zhu, J.J. Hull et al. 2016. Transcriptome analysis reveals a comprehensive insect resistance response mechanism in cotton to infestation by the phloem feeding insect *Bemisia tabaci* (whitefly). Plant Biotechnol J. 14: 1956–1975.

Li, Z.Q., S. Zhang, J.Y. Luo et al. 2013a. Ecological adaption analysis of the cotton aphid (*Aphis gossypii*) in different phenotypes by transcriptome comparison. PLoS One. 8: e83180.

Li, Z.Q., S. Zhang, Y. Ma et al. 2013b. First transcriptome and digital gene expression analysis in *Neuroptera* with an emphasis on chemoreception genes in *Chrysopa pallens* (Rambur). PLoS One. 8: e67151.

Li, Z.Q., S. Zhang, J.Y. Luo et al. 2015a. Transcriptome comparison of the sex pheromone glands from two sibling *Helicoverpa* species with opposite sex pheromone components. Sci Rep. 5: 9324.

Li, Z.Q., S. Zhang, J.Y. Luo et al. 2015b. Identification and expression pattern of candidate olfactory genes in *Chrysoperla sinica* by antennal transcriptome analysis. Comp Biochem Physiol Part D Genomics Proteomics. 15: 28–38.

Liu, N.Y., W. Xu, A. Papanicolaou et al. 2014. Identification and characterization of three chemosensory receptor families in the cotton bollworm *Helicoverpa armigera*. BMC Genomics. 15: 597.

Luo, J., X. Liu, L. Liu et al. 2014. *De novo* analysis of the *Adelphocoris suturalis* Jakovlev metathoracic scent glands transcriptome and expression patterns of pheromone biosynthesis-related genes. Gene. 551: 271–278.

Moreno, M. and E. Giralt. 2015. Three valuable peptides from bee and wasp venoms for therapeutic and biotechnological use: melittin, apamin and mastoparan. Toxins. 7: 1126–1150.

Offor, U.S., S. Nwi-Ue, A. Waka et al. 2014. Local methods of insect pest control in Ogoni lands rivers state—a review. Researcher. 6: 73–76.

Pan, Y., T. Peng, X. Gao et al. 2015a. Transcriptomic comparison of thiamethoxam-resistance adaptation in resistant and susceptible strains of *Aphis gossypii* Glover. Comp Biochem Physiol Part D Genomics Proteomics. 3: 10–15.

Pan, Y., C. Yang, X. Gao et al. 2015b. Spirotetramat resistance adaption analysis of *Aphis gossypii* Glover by transcriptomic survey. Pestic Biochem Physiol. 124: 73–80.

Pare, P.W. and J.H. Tumlinson. 1997. *De Novo* biosynthesis of volatiles induced by insect herbivory in cotton plants. Plant Physiol. 114: 1161–1167.

Peng, T., Y. Pan, C. Yang et al. 2016. Over-expression of CYP6A2 is associated with spirotetramat resistance and cross-resistance in the resistant strain of *Aphis gossypii* Glover. Pestic Biochem Physiol. 126: 64–69.

Poivet, E., A. Gallot, N. Montagné et al. 2013. A comparison of the olfactory gene repertoires of adults and larvae in the noctuid moth *Spodoptera littoralis*. PLoS One. 8: e60263.

Qiu, L., L. Hou, B. Zhang et al. 2015. Cadherin is involved in the action of *Bacillus thuringiensis* toxins Cry1Ac and Cry2Aa in the beet armyworm, *Spodoptera exigua*. J Invert Path. 127: 47–53.

Rangarirai, M., C. Blessing and M. Nhamo. 2015. Strategies for integrated management of co complex in Zimbabwe: A review. Int J Agron Agri Res. 7: 23–35.

Salvador, R., D. Príncipi and M. Berretta. 2014. Transcriptomic survey of the midgut of *Anthonomus grandis* (Coleoptera: Curculionidae). J Insect Sci. 14: 219.

Shaina, H., Z. UlAbdin, B.A. Webb et al. 2016. *De novo* sequencing and transcriptome analysis of venom glands of endoparasitoid *Aenasius arizonensis* (Girault) (=*Aenasius bambawalei* Hayat) (Hymenoptera, Encyrtidae). Toxicon. 121: 134–144.

Tao, X.Y., X.Y. Xue, Y.P. Huang et al. 2012. Gossypol-enhanced P450 gene pool contributes to cotton bollworm tolerance to a pyrethroid insecticide. Mol Ecol. 21: 4371–4385.

Tassone, E.E., G. Zastrow-Hayes, J. Mathis et al. 2016. Sequencing, *de novo* assembly and annotation of a pink bollworm larval midgut transcriptome. Gigasci. 5: 28.

Tian, C., W. Tek Tay, H. Feng et al. 2015. Characterization of *Adelphocoris suturalis* (Hemiptera: Miridae) transcriptome from different developmental stages. Sci Rep. 5: 11042.

Wang, J.L., L. Chen, L. Tang et al. 2014. 20-hydroxyecdysone transcriptionally regulates humoral immunity in the fat body of *Helicoverpa armigera*. Insect Mol Biol. 23: 842–856.

Wang, Q., A.E. Eneji, X. Kong et al. 2015. Salt stress effects on secondary metabolites of cotton in relation to gene expression responsible for aphid development. PLoS One. 10: e0129541.

Wu, X.M., C.Q. Yang, Y.B. Mao et al. 2016. Targeting insect mitochondrial complex I for plant protection. Plant Biotechnol J. 14: 1925–1935.

Xiong, G.H., L.S. Xing, Z. Lin et al. 2015. High throughput profiling of the cotton bollworm *Helicoverpa armigera* immunotranscriptome during the fungal and bacterial infections. BMC Genomics. 16: 321.

Xu, W., A. Papanicolaou, H.J. Zhang et al. 2016. Expansion of a bitter taste receptor family in a polyphagous insect herbivore. Sci Rep. 6: 23666.

Yong-Biao, L., B.E. Tabashnik, T.J. Dennehy et al. 1999. Development time and resistance to Bt crops. Nature. 400: 519.

Zou, D., T.A. Coudron, C. Liu et al. 2013. Nutrigenomics in *Arma chinensis*: transcriptome analysis of *Arma chinensis* fed on artificial diet and Chinese oak silk moth *Antheraea pernyi* pupae. PLoS One. 8: e60881.

CHAPTER 4

The Coevolution of the Plant-Insect Interaction Networks

Kevin Corneal, Jennifer Campbell, Nicholas Evans and *Chandrakanth Emani**

INTRODUCTION

Coevolution as a phenomenon had its foundations laid down by the pioneers of the theory of evolution. Charles Darwin while outlining the importance of natural selection (Darwin 1859) also emphasized that selective pressures in nature are the outcomes of interactions between organisms (Darwin 1862), an idea confirmed by Alfred Russell Wallace (Wallace 1889). In fact, Darwin's illustration of the idea involved the interaction between Orchids and insects (Darwin 1862). Early proponents of the phenomenon (without actually using the term coevolution) hypothesized that the apparent narrow host preferences of some insect herbivores might be explained by parallel evolution of insect clades and their host plants, and there are behavioral preferences of insects towards their host plants (Brues 1924). The term "coevolution" itself was first introduced to a broader audience in a seminal paper (Ehrlich and Raven 1964) through a comprehensive study of butterflies and their host plants, hypothesizing that the evolution of plant defenses followed by counter-adaptations in insects could lead to bursts of adaptive radiation. Subsequent studies suggested that such plant-insect interactions in nature are actually pollinator-driven speciation events that could be an important source of diversity in plant species (Grant and Grant 1965).

Department of Biology, Western Kentucky University-Owensboro, 4821 New Hartford Road, Owensboro, KY 42303.
* Corresponding author: chandrakanth.emani@wku.edu

Plant-insect interaction networks can thus be important illustrative examples of the phenomenon of coevolution in a dynamic environment where insect herbivores are under selection pressure to find quality hosts through complex interactive patterns that have evolved over millions of years though individual events can occur in milliseconds (Bruce 2015). Built into this interesting phenomenon is the crucial process of the insect fitness that involves finding ideal plant hosts while avoiding unsuitable ones and this is a result of the evolution of a finely tuned sensory system that detects host cues, and a nervous system that integrates inputs with complex spatio-temporal resolution (Bruce 2015). The myriad insect responses to cues are contextual and this involves changes in the physiological state of the insect and a well-observable display of prior learning experiences, where insects make and learn from 'mistakes' of being attracted to poor quality hosts and in a parallel display of mutual biological "arms race", plants undergo a selection pressure in the form of a detection evasion by defending themselves when attacked (Bruce 2015; Zheng and Dicke 2015). The so-called "arms race" between plant defenses and insect herbivore attackers involves insect-associated molecules that may trigger or suppress defence depending on whether the plant or the insect is ahead in evolutionary terms (Schoonhoven et al. 2005). Plant emit volatile compounds that are part of defence response mechanisms that are actually induced by insect feeding or oviposition which can attract natural enemies but repel herbivores (Will et al. 2007; Kessler and Baldwin 2002). In the frame of reference of the plants, this complex process results in changes related to plant reproductive fitness that is increased by attraction of pollinators in the form of diverse interactions that can be altered by other organisms associated with the plant such as other insects, plant pathogens, or mycorrhizal fungi (Bruce 2015).

Coevolution also involves the phenomenon of "mutualism" that is defined as the cooperative interactions between species and this played a central role in the generation and maintenance of life on earth (Bronstein et al. 2006). Plant insect–mutualism are manifest in the processes of pollination, protection and seed dispersal, and are inherent in five central phenomena: (a) the evolutionary origins and maintenance of mutualism, where the mutualistic interactions arose with plants evolving mechanisms that took advantage of insects foraging on plant tissues; (b) the evolution of mutualistic traits where, plants have undergone extensive trait evolution in the context of their mutualistic interactions with insects; (c) evolution of specialization (in terms of how certain plant species have on or few insect mutualists) and generalization (as in nutritional benefits for insects and plant protection mechanisms); (d) coevolutionary and cospeciation processes; and (e) the ubiquitous existence of cheating in terms of the foraging strategies employed by the insects that do not confer the advantages of pollination and seed dispersal to plants (extensively reviewed in Bronstein et al. 2006).

The Ecological and Evolutional Niche of Plant-Insect Interactions

Time and space have played a significant role in plant-insect interactions and recent research has shed light on the mechanistic basis by which insects interact with their host plants in nature (Hogenhout and Bos 2011; Mithoefer and Boland 2012; Smith and Clement 2012). The dynamic nature of the plant-insect interactions mandates exploration in both ecological and evolutionary realms as what is observed in on context may not repeat in another. Further, insect responses depend on diverse patterns of host cues with certain insects finding specific host plant species on which they can feed and reproduce as also certain non-host plant species that do not support feeding and/or reproduction of the insects pointing to a selection pressure to find quality hosts (Bruce et al. 2005). The process of locating suitable plants and avoid unsuitable hosts thus relates to a maximization of fitness in terms of the insects (Bruce and Pickett 2011). Insects have achieved this fitness in nature by evolving a finely tuned sensory system for detection of host cues and a nervous system capable of integrating inputs from sensory neurons with a high level of spatio-temporal resolution (Martin et al. 2011). In terms of time scales associated with plant-insect interactions, the phytophagous insects that exist today and the plants they feed on are the product of a coevolutionary process that has been ongoing for 400 million years (Labandeira 2013). The actual mechanisms themselves operate over much shorter periods, specifically, host plant decisions made in flight happen in milliseconds, settlement/colonization happening in minutes and induced defence mechanisms occurring in hours (Bruce 2014; Ton et al. 2007; Jinwon et al. 2011). The rapid insect responses to host plant cues from their external environment is mainly attributed to the sophisticated system for sensing their external environment and processing the sensory input (Martin et al. 2011). The plumes of insects are patchy distributions of high concentration packets of odor interspersed with clean air and are not smooth distributions of odor intensity (Cardé and Willis 2008). Insects shape their plume tracking behavior by making the exceedingly rapid decisions made during flight in a timescale of tens to hundreds of milliseconds using the wind direction bearing the odor as the primary directional cue that enables them to steer their movements toward the odor source (Cardé and Willis 2008). The olfactory part of the systems that are considered the "odor plume fluxes" (the odor quality in the insect brain) involves the activities of an array of thousands of tightly and differentially tuned olfactory receptor neurons

(ORNs) on the male antenna (Baker 2009). The ORNs have the distinct ability to recognize individual molecular structure of a blend of plant volatile compounds and this plays a pivotal role in both recognizing the correct host and avoiding non-hosts (Bruce and Pickett 2011). Additionally, the ability to recognize the right mix of plant volatiles confers a flexibility that allows for adaptating to variable environments by altered signal processing events while maintaining the same peripheral olfactory receptors (Bruce and Pickett 2011).

Studies involving the characterization and coding of behaviorally significant odor blends of plant volatiles showed elicitation of completely different insect responses as compared to exposure to individual compounds that highlight the fact that insect responses are sensitive to combinations of host cues (Riffell et al. 2009). As shown in the study conducted on black bean aphid, *Aphis fabae* and its host *Vicia faba*, insects showed a positive response to host volatile compounds when encountered together in a blend of 15 volatile compounds, but avoided the same volatiles when encountered individually (Webster et al. 2010). When a particular cue or set of cues are associated with a food reward, insect responses can also change with learning behavior (Hartleib et al. 1999).

Phytophagous insects as pests in agricultural ecosystems have a significant effect on food security for humanity (Bruce 2010), but other plant-insect interactions that are crucial to understand are related to the role of insects as pollinators combined with examining the characteristics of carnivorous plants that consume insects (Renner and Specht 2013). A comprehensive research encompassing all these natural events will be crucial for nature conservation in wild habitats, because the coevolutionary forces involved in the diverse plant-insect interactions can drive speciation and increase natural biodiversity.

Plant-Insect Interactions Viewed Through the Coevolution Paradigm

The evolutionary aspects of plant-insect interactions show crucial insights into the phenomenon of co-evolution. Evidence through fossil records date the origins of pollination to 250 million years (Labandeira 2013). This finding can be integrated into the fact that the diverse array of flowering plant species numbering over 275,000 are a direct result of recent adaptive radiation driven by the coevolution between plants and their beneficial animal pollinators (Yuan et al. 2013). Olfactory messages that posit specific plants to specific pollinators clearly suggest that a coevolution occurred between flowering plants and the related pollinators (Grajales-Conesa et al. 2011). This was clearly seen in orchids that mimic aphid alarm pheromones to attract hoverflies for pollination (Stoekl et al. 2011). In more recent studies, insect herbivores were shown to be major factors affecting the ecology and evolution of plants when suppressed insect attack in the fields of *Oenthora biennis* reduced seed predation and altered interspecific competitive dynamics in terms of declined resistance to herbivores due to changes in flowering time while increasing plant competitive ability (Aggarwal et al. 2012). This evidenced a real-time ecological and evolutionary change in plant populations that are interacting with insects and that natural selection favored different plant genotypes in the absence of herbivores rather than in their presence (Hare 2012). A causal link was also seen between variation in abundance of specialist insect herbivores and the geographic pattern of a polymorphic plant defense locus as seen in aphid-Arabidopsis interactions that points to a potency of insect herbivores as selection forces (Züst et al. 2012). Ecogenomic studies conducted in model plant Arabidopsis showed that coevolution was clearly at play between phytophagous insects adapting to exploit their hosts and the plants parallelly evolving defense mechanisms to counteract insect herbivory (Anderson and Mitchell-Olds 2011; Johnson 2011).

Plant insect interactions examined through a paleobiologic perspective evidence that insect herbivory can be dated to 400 million years (Labandeira 2013). Classic reviews such as those examining interactions between butterflies and plants (Erhlich and Raven 1964) clearly point to coevolution as the clear paradigm to understand community evolution where evolutionary interactions take place between organisms with minimal or no exchange of genetic interaction. Such ecological interactions are evolutionary conserved across the tree of life resulting in fitness advantages suggesting a shared pattern in the organization of biological systems mediated by conserved ecological interactions among taxa (Gomez et al. 2010).

Both field data and lab experimental studies examining host-parasite interactions (Benmayor et al. 2008) showed that coevolutionary interactions over time generate diversity by producing spatially divergent selection trajectories in interacting organisms (Laine 2009). More recent studies validated the hypothesis that organisms within trophic webs exert mutual selective pressures and the related biological interactions also promote genetic polymorphisms (Duffy and Forde 2009; Brown and Tellier 2011) that lead to quantifiable changes in gene frequency (Brockhurst and Koskella 2013). This offers conclusive proof that coevolution is an ecological phenomenon that also effects biological processes associated with genetic diversity. However, due caution needs to be ascertained in drawing conclusions as sometimes coadapted species do not have to be coevolved as seen in cases of interacting organisms simply exhibit pre-adaptations to persist in a community, the phenomenon termed as 'ecological fitting' (Janzen 1985). Invasive plants often integrate into native plant-pollinator networks without coevolving with native pollinators (Vila et al. 2009). Phytophagous insects can specialize in specific host plant use over time in spite of the fact that many agricultural pest species being polyphagous and this involves subtle and complex interplay between species involving multitrophic ecological interactions (Forister et al. 2012). An oscillation hypothesis was proposed in studies conducted with leaf-mining fly genus *Phytomyza* (Janz and Nylin 2008) to explain the periods of host range expansion followed by periods of specialization.

Genome level genetic divergence results due to divergent selection effects on ecological traits that may result in adaptive population differentiation and reproductive isolation were revealed in genome scans of pea aphid complex that specifically showed candidate regions involved in adaptation to host plant (Jaquiery et al. 2012). More specifically, differences between races were observed in olfactory receptor genes and three genes encoding salivary proteins of the pea aphid (Jaquiery et al. 2012), although it was unclear as to the exact timing of speciation when the related gene changes occurred or as to their causal role in the process of speciation. Another interesting observation was seen in *Drosophila sechellia* that evolved to specialize on *Morinda citrifolia* fruit (in comparison to the common fruit fly *Drosophila melanogaster*), where higher expression levels of neurons ab3 and ab3B render it sensitive to hexanoate esters and 2-heptanone, respectively enabling it to better recognize *Morinda* fruit odours (Ibba et al. 2010). All the observations discussed suggest the importance of time during which the coevolution process actually occurs especially in terms of the specific interactions that drive the evolution of novel species that maybe relatively brief, followed by the spread and usage of the resultant traits in novel environments (Suchan and Alvarez 2015).

Plant Volatile Organic Compounds and their Role in Odor Recognition

Insects exhibit complex mechanisms in the usage of plant volatiles to recognize their host plants in a specific host location (Bruce et al. 2005) and the processes also involve blends of commonly occurring plant volatiles in specific combinations or ratios, specifically in terms of finding the right mix of the plant volatile blends (Bruce and Pickett 2011). The dimension of the time in the host odor recognition seems to be a significant factor as the occurrence or lack of the simultaneous arrival of odors at the insect antenna can alter the type of the elicited behavioral response in the insect (Bruce 2015). The right blend of plant volatiles have a crucial role as evidenced in black bean aphid, *Aphis fabae*, where volatile blends that were effective host cues were rendered non-host cues when individual odors were presented in an olfactometer (Webster et al. 2010). This demonstrates the fact that the behavioral response not only depends on the molecular structure of the plant volatile but also on the context in which it is perceived (Bruce 2015). A 'coincidence detection' mechanism was suggested where locating a host plant is crucial for the resolution of odors and this allows insect herbivores to not only recognize host odor blends but to differentiate them from blends of non-host odors (Bruce et al. 2005). As seen in the tea aphid, *Toxoptera auratii*, combination of olfactory (plant volatile blends) and visual cues (colors of the tea plants) can further enhance attraction (Han et al. 2012). In terms of finding the right mix of plant volatiles, insects also exhibit active avoidance of non-host odors (Bruce and Pickett 2011). This was evidenced from earlier studies in the black bean aphid (Nottingham et al. 1991) where both behavioral and electrophysiological response observations showed the effective tuning of olfactory receptor neurones (ORNs) to specific non-host compounds such as 3-butenyl isothiocyanate and 4-pentenyl isothiocyanate. The isothiocynates tested in an olfactometer bioassay were found to be repellent. The significance of the ratios and concentration of the plant volatile compounds was also found to be crucial as seen in wind tunnel studies conducted in *Paralobesia viteana* (grape berry moth) (Cha et al. 2011) when it was observed that doubling the concentration of any one of the components of a synthetic host volatile blend of grape odors (comprising (*E*)- and (*Z*)-linalool oxides, nonanal, decanal, (*E*)-caryophyllene, and germacrene-D), while keeping the concentration of the other compounds constant, significantly reduced female attraction to host grape plants.

Evolution of Environmental and Developmental Changes in Insect Response

The plant defence signaling induced by insect herbivore attacks depends heavily on the insects' nervous system that confers on them a unique natural ability to learn and that has significant consequences for their responses to plant volatiles (Cunningham et al. 2004; Bruce and Pickett 2011). In terms of the innate and learned behavioral responses to odors in specific insect host locations, many insect species show a preference for plant volatiles experienced during development in the natal habitat (Webster et al. 2013). In feeding studies of black bean aphids, *Aphis fabae*, that were reared on artificial diets of natural host black bean and unsuitable host mustard, it was seen that insects preferred the leaves whose odor they had experienced during development, but in absence of cues of proximity and access, they preferred the natural bean volatiles. This points to the fact that information gained in natal habitat is mostly utilized in situations where the habitat exploration costs are the lowest (Webster et al. 2013). Learning behaviors that are associated with a reward were seen to affect both the strength and the type of response to plant stimuli as seen in hawkmoths (*Manduca sexta*) that are innately attracted to volatile blends of specific night-blooming flowers and in the absence of hawkmoth-adapted flowers in the habitat associate the odors of bat-pollinated *Agave palmeri* flowers which have a completely different smell (Riffell et al. 2013). The two examples discussed show that processing of stimuli through two olfactory channels involving both an innate bias and a learned association enables insect herbivores to successfully exist in constantly changing environments.

Insect responses can also be characterized as appetitive that is a natural response to satisfy developmental needs as illustrated by *Spodoptera littoralis* moths that were trained to extend their proboscis as a specific feeding response to (*Z*,*E*)-9,11-tetradecadienyl acetate that is a sex pheromone that usually elicits sexual behaviours (Hartlieb et al. 1999). An example of learned responses to odors as pertaining to behavioral discrimination of plant volatiles was seen in honey bees that learn to respond to linalool and 2-phenylethanol better than other oilseed rape volatiles (Pham-Delegue et al. 1993). This points to a hierarchy and an innate preference of insect herbivores for certain odors, specifically in a quest to find the right mix of plant volatile blends (Bruce and Pickett 2011).

Innate (natural or inborn) response to herbivore attack induced plant volatiles is usually seen in case of egg and larval parasitoids and specialists rely more on associative learning as seen in oviposition studies conducted with *Spodoptera frugiperda* in maize (Peñaflor et al. 2011a). The observed innate responses allow insects to respond rapidly to cues that occur in favorable environmental situations as illustrated by the oriental fruitfly (*Bactrosera dorsalis*) that responds to an oviposition stimulant, γ-octalactone, through an innate recognition template (IRT) mechanism (Damodaram et al. 2014) that can also be seen as an effective method to avoid detrimental environmental situations. A more effective example of the avoidance of unfavorable environments is the effect seen in *Drosophila melanogaster* where the insect has developed a specific conserved and dedicated olfactory circuit that respond exclusively to geosmin, a compound associated with harmful toxic microbes (Stensmyr et al. 2012).

Physiological condition of an insect, specifically as seen in hunger modulation studies in fruit flies (Ruebenbauer et al. 2008; Becher et al. 2010) significantly influences insect-plant interactions. Mating processes were also seen to induce significant physiological changes in female insects leading to behavioral adjustments to match the internal state of the animal as seen in the case of *S. littoralis* that switches its olfactory response from food to egg-laying cues following mating (Saveer et al. 2012). Calcium imaging using authentic and synthetic odors showed that unmated females that are strongly attracted to lilac flowers switch their attraction to the green-leaf odor of the larval host plant cotton (*Gossypium hirsutum*) due to the abolition of the floral odor, a 'floral to green switch' (Saveer et al. 2012).

Plant Defence Mechanisms

Plants have evolved diverse and comprehensive defence systems to protect their tissues from insect herbivores. The complex defence systems demonstrate a diverse array of herbivore induced behaviors and the research focus was on understanding the role of chemical cues in mediating the tritrophic (the plant, the herbivore and the natural enemies or predators of the herbivore) interactions (De Moraes et al. 2001). Chemical and behavioral assays in tobacco showed that the plants release temporally different volatile blends and the caterpillar herbivores use induced plant signals to choose sites for oviposition (De Moraes et al. 2001). Tobacco plants also effectively illustrated the fact that indirect defenses operate in plant defence processes when it was shown in studies at both field and lab levels that specific plant volatiles increased egg predation rates of generalist predators and volatile blends decreased oviposition rates (Kessler and Baldwin 2001). Expression of specific plant volatiles such as the oxylipn jasmonate is tightly regulated by ecological context of the plant through molecular signals that indicate the nature of the attacker, the value of the attacked organs, phytochrome status related to the proximity of competing plants, the natural associations with beneficial organisms and the history of the plant-pathogen-herbivore interactions (Ballare 2011). Among the plant volatiles, toxic or anti-feedant secondary metabolites were crucial barriers to herbivory (Harborne 1993) and were useful to recognize herbivory-associated molecular patterns involving signal perception and downstream signaling pathways involved in plant defence activation (Mithoefer and Boland 2012). Anatomical and chemical defenses of conifer bark against bark beetles involved physical defences such as lignin (Franceschi et al. 2005). Defense compounds also involved secondary metabolites such as protease inhibitors that were shown to be effective against the cotton bollworm, *Helicoverpa armigera* infecting wild pigeonpea plants (Parde et al. 2012), threonine deaminase in tomato degrading threonine in the insect gut (Gonzales-Vigil et al. 2011), 7-epizingiberene effecting the glandular trichomes of wild tomato (Bleeker et al. 2012), and *O*-acyl sugars in the glandular trichomes of tomato and other plants in the Solanaceae (Schilmiller et al. 2012).

Plant volatiles are involved in both constitutive as well as after attack induction. Such specific modes of action are illustrated by volatiles such as salicylic acid (SA) that is majorly associated with induced defence against pathogens while jasmonic acid (JA) is specific to herbivore defence (Ballare

2011). Recent studies have highlighted a complex paradigm of a "tripartite" interactions of plant-pathogen-arthropod involving both SA and JA pathways (Stout et al. 2006). Induced defence responses were also found to be elicited by biotic stresses that are inherent in herbivore and pathogen attack (Bruce and Pickett 2007). A diverse array of lepidopteran elicitors involving SA, JA and ethylene interactions mediated a biological 'cross-talk' in herbivory induced phytohormone signaling (Diezel et al. 2009).

Indirect defence mechanisms also involve plant secondary metabolism where attraction of natural enemies of pests such as parasitic wasps (Turlings et al. 1990) and caterpillars (De Moraes et al. 1998) point to multitrophic effects in an evolutionary context (Dicke and van Loon 2000), thus highlighting the tritrophic interaction scenario (Heil 2008).

Mutant studies investigating the wound-response pathways involved aphid-induced plant volatiles in the model plant *Arabidopsis thaliana* demonstrated that herbivore-induced plant volatile (HIPV) release specifically involves JA-signalling pathway (Girling et al. 2008). The findings were further corroborated by studies with *Myzus persicae* (green peach aphid) salivary component induced defence responses (de Vos and Jander 2009) and in tomato exposed to hawkmoth larvae where JA along with system in regulated the constitutive and herbivore-induced systemic volatile emissions (Degenhardt et al. 2010) but other systems could be different. In examples from other plant families, HIPVs included homoterpenes as seen in angiosperms that were seen as common constituents of floral and herbivore-induced plant volatiles (Tholl et al. 2011). HIPVs were also seen to increase plant fitness through improved bud and flower production in tobacco (Schuman et al. 2012). HIPV signatures in plants were seen to have high genetic variability in maize inbred lines (Degen et al. 2004) that affects a differential parasitism of herbivores on the inbred lines (Degen et al. 2012) and egg deposition by a herbivore was seen to induce the recruiting of egg and larval parasitoids (Tamiru et al. 2011).

HIPV emission can enable herbivores to successfully repel future colonization (de Moraes et al. 2001; Kessler and Baldwin 2001; Bruce et al. 2010). Maize HIPVs were seen to strongly influence the *Spodoptera frugiperda* moths (Signoretti et al. 2012). The preference for undamaged plants illustrates an effective ecological and evolutionary adaptive strategy aimed at tackling competitors and natural enemies for offspring (Bruce 2015).

Transcriptome and signal cascade analyses of volatile exposed plants revealed that plants "eavesdrop" to prime direct and indirect defenses and naturally induce their competitive abilities (Baldwin et al. 2006). Distance limitations were found to exist on neighbor plant signaling by airborne volatiles and it was observed that responses to HIPVs were observed to occur over relatively short distances (Frost et al. 2008) pointing to an adaptive mechanism to avoid a defence response unless concentrations are high enough to indicate a real threat (Bruce 2015).

Plants also exhibit an 'early herbivore alert' by induction of defence response to insect egg laying that is actually the very earliest stage of insect attack (Hilker and Meiners 2006). This sort of priming has a very high adaptive value as the plant is developing the defence mechanism pathways much ahead of the more damaging feeding stages of the insect life cycle. The recruiting of egg and larval parasitoids in maize in response to herbivore egg deposition (Tamiru et al. 2011) and the attraction of parasitic wasps by *Brassica nigra* volatiles induced by oviposition of the cabbage butterfly (Fatouros et al. 2012) are illustrations of this scenario. Plants exposed to herbivore eggs also were found to increase direct defences to curtail insect growth rates thus acting as a warning signal to ward off future feeding damage (Beyaert et al. 2012; Jinwon et al. 2012; Geiselhardt et al. 2013).

Evolutionary Changes in Plant Responses

Plant defence against insect herbivores involving both direct and indirect chemical defences result in highly complex molecular regulatory networks that lead to gene activation by signaling pathways that constitute reactive oxygen species and calcium signatures (Maffei et al. 2007). Both constitutive or induced defences crucially involve the dimension of time in the most important event of priming of defence as a way of preparing for the battle against insect attacks (Conrath et al. 2006) involving air-borne signals (Ton et al. 2007) and biotic stress (Bruce et al. 2007). Primed plants respond more quickly and strongly upon subsequent attacks (Ton et al. 2007) and the induced resistance resulting from priming viewed in the evolutionary paradigm actually represents the plants ability to integrate multiple suites of signals related to the evolutionary origins of bot the plant and the herbivore (Jinwon et al. 2011). In their 350 million period of evolutionary coexistence, the plants and insects have evolved a variety of different interactions and one of the significant evolutionary innovation has been that of the production of herbivore-derived metabolites that were efficiently allocated to prime and induce defensive activities in precise timing to result in adaptive modulations of the plants' metabolism (Mithoefer and Boland 2012).

The early events in the molecular signaling pathways result in opening calcium channels and trigerring a series of cascade events, including reactive oxygen species (ROS) production (Maffei et al. 2007). The calcium channels were found to be associated with plant receptors that are chemically tuned to insect elicitors and rapid increases in ROS concentration was found to follow tissue damage caused by both biotic and abiotic injuries. Herbivore-associated wounding is different from mechanical wounding in terms of the Ca2+ influx and depolarization being maintained after herbivore wounding unlike mechanical wounding (Maffei et al. 2004). Membrane depolarization that follow leaf wounding results in the systemic spread of herbivore-induced defence through a plant that was correlated to JA-signaling domains resulting in a transcriptome enriched in RNAs encoding key jasmonate signal regulators (Mousavi et al. 2013). Epigenetic imprinting can lead to long-term evolutionary changes after biotic stress (Bruce et al. 2007) leading to a next-generation systemic acquired resistance (Luna et al. 2012) pointing to the fact that herbivory in a previous generation actually primes subsequent plant generations for increased insect resistance (Rasmann et al. 2012).

Insights gained from molecular modeling studies demonstrated that insects in the course of their evolution can overcome plant induced defenses by coevolving adaptations such as cytochrome P450 monooxygenases (P450s) that metabolize plant toxins (Schuler and Berenbaum 2013). This points to the fact that coevolution has a significant effect in changing the biological role of plant defence chemicals over evolutionary time scales. Illustrations of such biological changes can be seen in *Helicoverpa armigera* (cotton bollworm) that uses a P450, CYP6AE14, to detoxify gossypol (Mao et al. 2007); hawkmoth feeding on *O*-acyl-sugar-producing *Nicotiana attenuata* (Weinhold and Baldwin 2011); and many Brassica specialists that have evolved adaptations to thrive on glucosinolate-producing plants (Winde and Wittstock 2011; Bruce 2014). A study on the pheromone biology in danaid butterflies' evolution in terms of their chemical communication, plant relationships and mimicry showcased a tritropic level event where the insects utilized the plant secondary metabolites to defend themselves against their own attackers (Boppré 1978). A case of convergent evolution involving a specific amino acid change on a transmembrane sodium channel targeting cardenolide toxins was observed in several insect species (Dobler et al. 2012). In a more recent study, gene amplification and microsatellite polymorphism were seen to be involved in a recent insect host shift involving the peach-potato aphid, *Myzus persicae* that attacks tobacco, where the tobacco-adapted aphid races were found to overexpress a cytochrome P450 enzyme (CYP6CY3) that allows them to detoxify nicotine (Bass et al. 2013).

The Biochemical Array for Insect Recognition by Plants

The response to chemical stimuli that forms an overarching part of the plant-insect interactions in nature involving the effective detection of molecules associated with herbivore attack is crucial in eliciting behavioral, physiological, and biochemical response in plants that ensures their survival in nature. Over millions of years spanning both plant and insect evolution, plants have evolved comprehensive methods to detect and ward off insect herbivores and this process involves the ability to respond to a wide range of biomolecules. A process that involves the perception of microbe-associated molecular patterns and danger signals by pattern recognition receptors is a fundamental to the immune responses of both plants and animals (Boller and Felix 2009) and this forms the basis for pathogen recognition and innate immunity (Akira et al. 2006). Molecular recognition via ligand–receptor binding phenomena plays a significant role in plants (Boller and Felix 2009) as illustrated in case of plant pattern recognition receptor complexes in plasma membranes (Monaghan and Zipfel 2012). The JA pathway is a major signaling biochemical cassette that effectively integrates information perceived at the plant-insect interaction interface to translate into broad-spectrum defense responses underlining the crucial role of phytohormones in insect-specific plant reactions (Erb et al. 2012). Innate immunity conferred to plants by leucine-rich receptors and cytochrome P450 related proteins (Prince et al. 2014) and conserved bacterial proteins in the insect endosymbionts (Chaudhary et al. 2014) add further proof to the importance of receptors and ligands as being crucial to understand specificity in plant immunity to herbivore. Over the evolutionary time scale, plants have developed surveillance systems that are able to detect highly specific herbivore-associated cues as well as general patterns of cellular damage that induce specific defences with molecular recognition mechanisms underpinning this process involving receptors tuned to herbivore-associated molecular patterns (HAMPs; Mithofer and Boland 2008; Bonaventure 2012). A damaged-self recognition based on plant-derived elicitors that induces octadecanoid signaling plays a significant role in plant sensing of wounds (Heil et al. 2012).

Conclusions

Plant-insect interactions viewed through the reference frames of evolution and ecology point to complex and dynamic biological phenomena. The biological events also need to be effectively examined in the dimension of time as snapshots of one phenomenon may not overlap with another. It should be noted that the evolutionary changes that occur in both the plant and insect realms cannot be viewed in isolation and the approach to examine their coevolution is a more unbiased research focus. In case of the insects, changes fall into the realm of learning behavior patterns in the short term and the genetic mutations' realm in the longer term. Plant changes on the other hand are more focused on induced defence processes in the short term, epigenetic changes in the medium term, and gene mutations in the longer term. The environment also plays a crucial role as it changes over time and this makes the study of interactions more complicated even further and proper testable conclusions will be made if we consider the ecological perspective.

References

Agrawal, A.A., A.P. Hastings, M.T. Johnson et al. 2012. Insect herbivores drive real-time ecological and evolutionary change in plant populations. Science. 338: 113–116.

Akira, S., S. Uematsu and O. Takeuchi. 2006. Pathogen recognition and innate immunity. Cell. 124: 783–801.

Anderson, J.T. and T. Mitchell-Olds. 2011. Ecological genetics and genomics of plant defences: evidence and approaches. Functional Ecol. 25: 312–324.

Arimura, G., K. Matsui and J. Takabayashi. 2009. Chemical and molecular ecology of herbivore-induced plant volatiles: proximate factors and their ultimate functions. Plant Cell Physiol. 50: 911–923.

Baker, T. 2009. Representations of odor plume flux are accentuated deep within the moth brain. J Biol. 8: 16.

Baldwin, I.T., R. Halitschke, A. Paschold et al. 2006. Volatile signaling in plant-plant interactions: "Talking trees" in the genomics era. Science. 311: 812–815.

Ballare, C.L. 2011. Jasmonate-induced defenses: a tale of intelligence, collaborators and rascals. Trends Plant Sci. 16: 249–257.

Bass, C., C.T. Zimmer, J.M. Riveron et al. 2013. Gene amplification and microsatellite polymorphism underlie a recent insect host shift. Proc Natl Acad Sci USA. 110: 19460–19465.

Benmayor, R., A. Buckling, M.B. Bonsall et al. 2008. The interactive effects of parasites, disturbance, and productivity on experimental adaptive radiations. Evol. 62: 467–477.

Beyaert, I., D. Koepke, J. Stiller et al. 2012. Can insect egg deposition 'warn' a plant of future feeding damage by herbivorous larvae? Proc Royal Soc B: Biol Sci. 279: 101–108.

Bleeker, P.M., R. Mirabella, P.J. Diergaarde et al. 2012. Improved herbivore resistance in cultivated tomato with the sesquiterpene biosynthetic pathway from a wild relative. Proc Natl Acad Sci USA. 109: 20124–20129.

Boller, T. and G. Felix. 2009. A renaissance of elicitors: Perception of microbe-associated molecular patterns and danger signals by pattern-recognition receptors. Annu Rev Plant Biol. 60: 379–406.

Bonaventure, G. 2012. Perception of insect feeding by plants. Plant Biol. 14: 872–880.

Boppré, M. 1978. Chemical communication, plant relationships, and mimicry in the evolution of danaid butterflies. Entomol Experimental Appl. 24: 264–277.

Bronstein, J.L., R. Alacron and M. Geber. 2006. The evolution of plant-insect mutualisms. New Phytol. 172: 412–428.

Brown, J.K.M. and A. Tellier. 2011. Plant-parasite coevolution: bridging the gap between genetics and ecology. Annu Rev Phytopathol. 49: 345–367.

Bruce, T. 2014. Glucosinolates in oilseed rape: secondary metabolites that influence interactions with herbivores and their natural enemies. Ann Appl Biol. 164: 348–353.

Bruce, T.J., L.J. Wadhams and C.M. Woodcock. 2005. Insect host location: a volatile situation. Trends Plant Sci. 10: 269–274.

Bruce, T.J., M.C. Matthes, J.A. Napier et al. 2007. Stressful memories of plants. Plant Sci. 173: 603–608.

Bruce, T.J., C.A. Midega, M.A. Birkett et al. 2010. Is quality more important than quantity? Insect behavioral responses to changes in a volatile blend after stemborer oviposition on an African grass. Biol Lett. 6: 314–317.

Bruce, T.J.A. and J.A. Pickett. 2007. Plant defence signaling induced by biotic attacks. Curr Opin Plant Biol. 10: 387–392.
Bruce, T.J.A. 2010. Tackling the threat to food security caused by crop pests in the new millennium. Food Sec. 2: 133–141.
Bruce, T.J.A. and J.A. Pickett. 2011. Perception of plant volatile blends by herbivorous insects—Finding the right mix. Phytochem. 72: 1605–1611.
Bruce, T.J.A. 2012. GM as a route for delivery of sustainable crop protection. J Expl Bot. 63: 537–541.
Bruce, T.J.A. 2015. Interplay between insects and plants: dynamic and complex interactions that have coevolved over millions of years but act in milliseconds. J Exp Bot. 66: 455–465.
Brues, C. 1924. The specificity of food-plants in the evolution of phytophagous insects. American Naturalist. 58: 127–144.
Cardé, R. and M. Willis. 2008. Navigational strategies used by insects to find distant, wind-borne sources of odor. J Chem Ecol. 34: 854–866.
Cha, D.H., C.E. Linn, Jr., P.E.A. Teal et al. 2011. Eavesdropping on plant volatiles by a specialist moth: Significance of ratio and concentration. PLoS ONE. 6: e17033.
Chaudhary, R., H.S. Atamian, Z. Shen et al. 2014. GroEL from the endosymbiont *Buchnera aphidicola* betrays the aphid by triggering plant defense. Proc Natl Acad Sci USA. 24: 8919–8924.
Conrath, U., G.J.M. Beckers, V. Flors et al. 2006. Priming: Getting ready for battle. MPMI. 19: 1062–1071.
Cunningham, J.P., C.J. Moore, M.P. Zalucki et al. 2004. Learning odour preference and flower foraging in moths. J Exp Biol. 207: 87–94.
Damodaram, K.J.P., V. Kempraj, R.M. Aurade et al. 2014. Oviposition site-selection by *Bactrocera dorsalis* is mediated through an innate recognition template tuned to γ-octalactone. PLoS ONE. 9: e85764.
Darwin, C. 1859. On the Origin of Species by Means of Natural Selection. Murray, London, UK.
Darwin, C. 1862. On the Various Contrivances by which British and Foreign Orchids are Fertilised by Insects: and on the Good Effects of Intercrossing. Murray, London, UK.
De Moraes, C.M., W.J. Lewis, P.W. Pare et al. 1998. Herbivore-infested plants selectively attract parasitoids. Nature. 393: 570–573.
De Moraes, C.M., M.C. Mescher and J.H. Tumlinson. 2001. Caterpillar induced nocturnal plant volatiles repel conspecific females. Nature. 410: 577–580.
De Vos, M. and G. Jander. 2009. *Myzus persicae* (green peach aphid) salivary components induce defence responses in *Arabidopsis thaliana*. Plant Cell Environ. 32: 1548–1560.
Degen, T., C. Dillmann, F. Marion-Poll et al. 2004. High genetic variability of herbivore-induced volatile emission within a broad range of maize inbred lines. Plant Physiol. 135: 1928–1938.
Degen, T., N. Bakalovic, D. Bergvinson et al. 2012. Differential performance and parasitism of caterpillars on maize inbred lines with distinctly different herbivore-induced volatile emissions. PLoS ONE. 7: e47589.
Degenhardt, D.C., S. Refi-Hind, J.W. Stratmann et al. 2010. Systemin and jasmonic acid regulate constitutive and herbivore induced systemic volatile emissions in tomato, *Solanum lycopersicum*. Phytochem. 71: 2024–2037.
Dicke, M. and J.J.A. van Loon. 2000. Multitrophic effects of herbivore-induced plant volatiles in an evolutionary context. Entomol Exp Appl. 97: 237–249.
Diezel, C., C.C. von Dahl, E. Gaquerel et al. 2009. Different Lepidopteran elicitors account for cross-talk in herbivory-induced phytohormone signaling. Plant Physiol. 150: 1576–1586.
Dobler, S., S. Dalla, V. Wagschal et al. 2012. Communitywide convergent evolution in insect adaptation to toxic cardenolides by substitutions in the Na,K-ATPase. Proc Natl Acad Sci USA. 109: 13040–13045.
Doss, R.P., J.E. Oliver, W.M. Proebsting et al. 2000. Bruchins: Insect derived plant regulators that stimulate neoplasm formation. Proc Natl Acad Sci USA. 97: 6218–6223.

Du, B., W. Zhang, B. Liu et al. 2009. Identification and characterization of Bph14, a gene conferring resistance to brown planthopper in rice. Proc Natl Acad Sci USA. 106: 22163–22168.

Duffy, M.A. and S.E. Forde. 2009. Ecological feedbacks and the evolution of resistance. J Animal Ecol. 78: 1106–1112.

Ehrlich, P.R. and P.H. Raven. 1964. Butterflies and plants: a study in co-evolution. Evolution. 18: 586–608.

Erb, M., S. Meldau and G.A. Howe. 2012. Role of phytohormones in insect-specific plant reactions. Trends Plant Sci. 17: 250–259.

Fatouros, N.E., C. Broekgaarden, G. Bukovinszkine'Kiss et al. 2008. Male-derived butterfly anti-aphrodisiac mediates induced indirect plant defense. Proc Natl Acad Sci USA. 105: 10033–10038.

Fatouros, N.E., D. Lucas-Barbosa, B.T. Weldegergis et al. 2012. Plant volatiles induced by herbivore egg deposition affect insects of different trophic levels. PLoS ONE. 7: e43607.

Forister, M.L., L.A. Dyer, M.S. Singer et al. 2012. Revisiting the evolution of ecological specialization, with emphasis on insect-plant interactions. Ecology. 93: 981–991.

Franceschi, V.R., P. Krokene, E. Christiansen et al. 2005. Anatomical and chemical defenses of conifer bark against bark beetles and other pests. New Phytol. 167: 353–375.

Frost, C.J., M.C. Mescher, J.E. Carlson et al. 2008. Why do distance limitations exist on plant-plant signaling via airborne volatiles? Plant Signal Beh. 3: 466–468.

Geiselhardt, S., K. Yoneya, B. Blenn et al. 2013. Egg laying of cabbage white butterfly (*Pieris brassicae*) on *Arabidopsis thaliana* affects subsequent performance of the larvae. PLoS ONE. 8: e59661.

Girling, R.D., R. Madison, M. Hassall et al. 2008. Investigations into plant biochemical wound-response pathways involved in the production of aphid-induced plant volatiles. J Exp Bot. 59: 3077–3085.

Gomez, J.M., M. Verdu and F. Perfectti. 2010. Ecological interactions are evolutionarily conserved across the entire tree of life. Nature. 465: 918–921.

Gonzales-Vigil, E., C.M. Bianchetti, G.N. Phillips, Jr. et al. 2011. Adaptive evolution of threonine deaminase in plant defense against insect herbivores. Proc Natl Acad Sci USA. 108: 5897–5902.

Grajales-Conesa, J., V. Melendez-Ramirez and L. Cruz-Lopez. 2011. Floral scents and their interaction with insect pollinators. Rev Mex Biodiv. 82: 1356–1367.

Grant, V. and K.A. Grant. 1965. Flower Pollination in the Phlox Family. Columbia University Press, New York, NY, USA.

Han, B., Q.-H. Zhang and J.A. Byers. 2012. Attraction of the tea aphid, *Toxoptera aurantii*, to combinations of volatiles and colors related to tea plants. Entomol Exper Appl. 144: 258–269.

Harborne, J.B. 1993. Introduction to Ecological Biochemistry. London: Academic press.

Hare, J.D. 2012. How insect herbivores drive the evolution of plants. Science. 338: 50–51.

Hartlieb, E., B.S. Hansson and P. Anderson. 1999. Sex or food? Appetetive learning of sex odors in a male moth. Naturwissenschaften. 86: 396–399.

Heil, M. 2008. Indirect defence via tritrophic interactions. New Phytol. 178: 41–61.

Heil, M., E. Ibarra-Laclette, R.M. Adame-Alvarez et al. 2012. How plants sense wounds: Damaged-self recognition is based on plant-derived elicitors and induces octadecanoid signaling. PLoS ONE. 7: e30537.

Hilker, M. and T. Meiners. 2006. Early herbivore alert: Insect eggs induce plant defense. J Chem Ecol. 32: 1379–1397.

Ibba, I., A.M. Angioy, B.S. Hansson et al. 2010. Macroglomeruli for fruit odors change blend preference in Drosophila. Naturwissenschaften. 97: 1059–1066.

Janz, N. and S. Nylin. 2008. The oscillation hypothesis of host plant-range and speciation. pp. 203–215. *In*: Tilmon, K.J. (ed.). Specialization, Speciation, and Radiation: The Evolutionary Biology of Herbivorous Insects. Berkeley, California: University of California Press.

Janzen, D. 1985. On ecological fitting. Oikos. 45: 308–310.
Jaquiery, J., S. Stoeckel, P. Nouhaud et al. 2012. Genome scans reveal candidate regions involved in the adaptation to host plant in the pea aphid complex. Molecular Ecol. 21: 5251–5264.
Jinwon, K., H. Quaghebeur and G.W. Felton. 2011. Reiterative and interruptive signaling in induced plant resistance to chewing insects. Phytochem. 72: 1624–1634.
Jinwon, K., J.F. Tooker, D.S. Luthe et al. 2012. Insect eggs can enhance wound response in plants: A study system of tomato *Solanum lycopersicum* L. and *Helicoverpa zea Boddie*. PLoS ONE. 7: e37420.
Johnson, M.T.J. 2011. Evolutionary ecology of plant defences against herbivores. Funct Ecol. 25: 305–311.
Kessler, A. and I.T. Baldwin. 2001. Defensive function of herbivore-induced plant volatile emissions in nature. Science. 291: 2141–2144.
Kessler, A. and I.T. Baldwin. 2002. Plant responses to insect herbivory: the emerging molecular analysis. Annu Rev Plant Biol. 53: 299–328.
Labandeira, C.C. 2013. A paleobiologic perspective on plant–insect interactions. Curr Opin Plant Biol. 16: 414–421.
Laine, A.-L. 2009. Role of coevolution in generating biological diversity: spatially divergent selection trajectories. J Exp Bot. 60: 2957–2970.
Lucas-Barbosa, D., J.J.A. van Loon and M. Dicke. 2011. The effects of herbivore-induced plant volatiles on interactions between plants and flower-visiting insects. Phytochem. 72: 1647–1654.
Luna, E., T.J. Bruce, M.R. Roberts et al. 2012. Next-generation systemic acquired resistance. Plant Physiol. 158: 844–853.
Maffei, M., S. Bossi, D. Spiteller et al. 2004. Effects of feeding *Spodoptera littoralis* on lima bean leaves. I. Membrane potentials, intracellular calcium variations, oral secretions, and regurgitate components. Plant Physiol. 134: 1752–1762.
Maffei, M.E., A. Mithofer and W. Boland. 2007. Insects feeding on plants: Rapid signals and responses preceding the induction of phytochemical release. Phytochem. 68: 2946–2959.
Mao, Y.B., W.J. Cai, J.W. Wang et al. 2007. Silencing a cotton bollworm P450 monooxygenase gene by plant-mediated RNAi impairs larval tolerance of gossypol. Nat Biotechnol. 25: 1307–1313.
Martin, J.P., A. Beyerlein, A.M. Dacks et al. 2011. The neurobiology of insect olfaction: Sensory processing in a comparative context. Prog Neurobiol. 95: 427–447.
Mithofer, A. and W. Boland. 2008. Recognition of herbivory-associated molecular patterns. Plant Physiol. 146: 825–831.
Mithoefer, A. and W. Boland. 2012. Plant defense against herbivores: Chemical aspects. Annu Rev Plant Biol. 63: 431–450.
Monaghan, J. and C. Zipfel. 2012. Plant pattern recognition receptor complexes at the plasma membrane. Curr Opin Plant Biol. 15: 349–357.
Mousavi, S.A.R., A. Chauvin, F. Pascaud et al. 2013. Glutamate receptor-like genes mediate leaf-to-leaf wound signalling. Nature. 500: 422–426.
Nottingham, S.F., J. Hardie, G.W. Dawson et al. 1991. Behavioral and electrophysiological responses of aphids to host and nonhost plant volatiles. J Chem Ecol. 17: 1231–1242.
Parde, V.D., H.C. Sharma and M.S. Kachole. 2012. Protease inhibitors in wild relatives of pigeonpea against the cotton bollworm/legume pod borer, *Helicoverpa armigera*. Amer J Plant Sci. 3: 627–635.
Peñaflor, M.F.G.V., M. Erb, L.A. Miranda et al. 2011. Oviposition by a moth suppresses constitutive and herbivore-induced plant volatiles in maize. Planta. 234: 207–215.
Pham-Delegue, M.H., O. Bailez, M.M. Blight et al. 1993. Behavioral discrimination of oilseed rape volatiles by the honeybee *Apis mellifera* L. Chem Senses. 18: 483–494.
Prince, D.C., C. Drurey, C. Zipfel et al. 2014. The leucinerich repeat receptor-like kinase brassinosteroid insensitive1-associated kinase1 and the cytochrome p450 phytoalexin

deficient3 contribute to innate immunity to aphids in Arabidopsis. Plant Physiol. 164: 2207–2219.

Rasmann, S., M. De Vos, C.L. Casteel et al. 2012. Herbivory in the previous generation primes plants for enhanced insect resistance. Plant Physiol. 158: 854–863.

Renner, T. and C.D. Specht. 2013. Inside the trap: gland morphologies, digestive enzymes, and the evolution of plant carnivory in the Caryophyllales. Curr Opin Plant Biol. 16: 436–442.

Riffell, J.A., H. Lei, T.A. Christensen et al. 2009. Characterization and coding of behaviorally significant odor mixtures. Curr Biol. 19: 335–340.

Riffell, J.A., H. Lei, L. Abrell et al. 2013. Neural basis of a pollinator's buffet: olfactory specialization and learning in *Manduca sexta*. Science. 339: 200–204.

Robert, C.A.M., M. Erb, M. Duployer et al. 2012. Herbivore-induced plant volatiles mediate host selection by a root herbivore. New Phytol. 194: 1061–1069.

Ruebenbauer, A., F. Schlyter, B.S. Hansson et al. 2008. Genetic variability and robustness of host odor preference in *Drosophila melanogaster*. Curr Biol. 18: 1438–1443.

Saveer, A.M., S.H. Kromann, G. Birgersson et al. 2012. Floral to green: mating switches moth olfactory coding and preference. Proc Royal Soc B: Biol Sci. 279: 2314–2322.

Schilmiller, A.L., A.L. Charbonneau and R.L. Last. 2012. Identification of a BAHD acetyltransferase that produces protective acyl sugars in tomato trichomes. Proc Natl Acad Sci USA. 109: 16377–16382.

Schoonhoven, L.M., J.J.A. van Loon and M. Dicke. 2005. Insect-plant Biology, Ed 2 Oxford University Press, Oxford.

Schuler, M.A. and M.R. Berenbaum. 2013. Structure and function of Cytochrome P450S in insect adaptation to natural and synthetic toxins: insights gained from molecular modeling. J Chem Ecol. 39: 1232–1245.

Schuman, M.C., K. Barthel and I.T. Baldwin. 2012. Herbivory-induced volatiles function as defenses increasing fitness of the native plant *Nicotiana attenuata* in nature. eLife. 1: e00007.

Signoretti, A.G.C., M.F.G.V. Peñaflor and J.M.S. Bento. 2012. Fall Armyworm, *Spodoptera frugiperda* (JE Smith) (Lepidoptera: Noctuidae), female moths respond to herbivore-induced corn volatiles. Neotrop Entomol. 41: 22–26.

Smith, C.M. and S.L. Clement. 2012. Molecular bases of plant resistance to arthropods. Annu Rev Entomol. 57: 309–328.

Stensmyr, M.C., H.K.M. Dweck, A. Farhan et al. 2012. A conserved dedicated olfactory circuit for detecting harmful microbes in Drosophila. Cell. 151: 1345–1357.

Stoekl, J., J. Brodmann, A. Dafni et al. 2011. Smells like aphids: orchid flowers mimic aphid alarm pheromones to attract hoverflies for pollination. Proc Royal Soc B: Biol Sci. 278: 1216–1222.

Stout, M.J., J.S. Thaler and B.P.H.J. Thomma. 2006. Plant-mediated interactions between pathogenic microorganisms and herbivorous arthropods. Annu Rev Entomol. 51: 663–689.

Tamiru, A., T.J.A. Bruce, C.M. Woodcock et al. 2011. Maize landraces recruit egg and larval parasitoids in response to egg deposition by a herbivore. Ecol Lett. 14: 1075–1083.

Tholl, D., R. Sohrabi, J.-H. Huh et al. 2011. The biochemistry of homoterpenes—Common constituents of floral and herbivore-induced plant volatile bouquets. Phytochem. 72: 1635–1646.

Ton, J., M. D'Alessandro, V. Jourdie et al. 2007. Priming by airborne signals boosts direct and indirect resistance in maize. Plant J. 49: 16–26.

Turlings, T.C.J., J.H. Tumlinson and W.J. Lewis. 1990. Exploitation of herbivore-induced plant odors by host-seeking parasitic wasps. Science. 250: 1251–1253.

Vila, M., I. Bartomeus, A.C. Dietzsch et al. 2009. Invasive plant integration into native plant pollinator networks across Europe. Proc Roy Soc Lond B. 276: 3887–3893.

Webster, B., T. Bruce, J. Pickett et al. 2010. Volatiles functioning as host cues in a blend become non-host cues when presented alone to the black bean aphid. Animal Beh. 79: 451–457.

Webster, B., E. Qvarfordt, U. Olsson et al. 2013. Different roles for innate and learnt behavioral responses to odors in insect host location. Behavioral Ecol. 24: 366–372.
Weinhold, A. and I.T. Baldwin. 2011. Trichome-derived O-acyl sugars are a first meal for caterpillars that tags them for predation. Proc Natl Acad Sci USA. 108: 7855–7859.
Winde, I. and U. Wittstock. 2011. Insect herbivore counteradaptations to the plant glucosinolate-myrosinase system. Phytochem. 72: 1566–1575.
Wu, J. and I.T. Baldwin. 2009. Herbivory-induced signaling in plants: perception and action. Plant Cell Environ. 32: 1161–1174.
Yuan, Y.-W., K.J.R.P. Byers and H.D. Bradshaw Jr. 2013. The genetic control of flower-pollinator specificity. Curr Opin Plant Biol. 16: 422–428.
Zheng, S. and M. Dicke. 2008. Ecological genomics of plant-insect interactions from gene to community. Plant Physiol. 146: 812–817.
Züst, T., C. Heichinger, U. Grossniklaus et al. 2012. Natural enemies drive geographic variation in plant defenses. Science. 338: 116–119.

CHAPTER 5

Linking Primary and Secondary Metabolism

A Mechanistic Hypothesis for how Elevated CO_2 Modulates Defenses

Linus Gog,[1] *Jorge Zavala*[2] and *Evan H. DeLucia*[1,*]

INTRODUCTION

Rising atmospheric CO_2 affects insect herbivory by altering both the primary and secondary metabolism of plants (Zavala et al. 2013). In terms of primary metabolism, the portion of plant physiology immediately concerned with growth, elevated CO_2 increases the rate of photosynthesis and thus the accumulation of starch in leaves. The consequences of this effect on insect feeding behavior are two-fold: With a more concentrated supply of carbohydrates, insect herbivores are provided with chemical energy to increase their rates of feeding (Lincoln et al. 1986; for review see Zavala et al. 2013). At the same time, increased starch content dilutes foliar nitrogen and thus obliges insect herbivores to compensate their nitrogen requirements by ingesting more leaf tissue (Lincoln et al. 1986; for review see Zavala et al. 2013). Meanwhile, the effects of elevated CO_2 on plant secondary metabolism—the portion of plant physiology responsible for mediating ecological interactions through chemical defense—are comparatively resistant to generalization. The source of uncertainty in understanding how elevated CO_2 influences insect feeding behavior rests in the regulatory connection between primary and secondary metabolism.

[1] Department of Plant Biology, University of Illinois at Urbana-Champaign, 265 Morrill Hall, 505 South Goodwin, Urbana, Il 61801.
Email: linusgog@illinois.edu

[2] Facultad de Agronomia, Universidad de Buenos Aires Cátedra de Bioquímica, Av. San Martín 4453 C141 DSE, Argentina, C1121 ABG.
Email: zavala@agro.uba.ar

* Corresponding author: delucia@illinois.edu

Soybean (*Glycine max*) grown under elevated CO_2 at the soyFACE field experiment exhibits a pattern of altered responses to common insect crop pests that has no apparent physiological connection to altered carbohydrate metabolism. Leaf herbivory was considerably greater on soybean grown under elevated than ambient CO_2, and increased damage was associate with larger populations of Japanese Beetles (*Papilio japonica*) and soybean aphid (*Aphis glycines*) (Hamilton et al. 2005; Dermody et al. 2008). Increased susceptibility of soybean under elevated CO_2 was not strictly related to increased leaf carbohydrates (O'Neill et al. 2008), but was instead caused primarily by a reduction in chemical defenses. Zavala et al. (2008) report that induction of cysteine protease inhibitors (CysPIs) is significantly slower in soybean grown under elevated CO_2 relative to control. These CysPIs render soybean tissue indigestible to insect herbivores by blocking proteases found in insect guts; they are especially important against coleopteran pests, whose digestive capacity is otherwise robust. Japanese Beetles are common to soybean fields in the American Midwest (Zavala et al. 2009), as are adults of Western Corn Rootworm, *Diabrotica virgifera*, of which a rotation-resistant strain is an increasing cause for concern (Levine et al. 2002). Beyond altered production of CysPIs, O'Neill et al. (2010) report that foliar concentration of the flavonoid quercetin is higher under elevated CO_2 relative to control. Insects feeding on high CO_2 soybean may derive a benefit from consuming this antioxidant, further stimulating herbivore damage. Similar studies by Guo et al. (2012) and Zhang et al. (2015) on tomato, Matros et al. (2006) on tobacco, and Mhamdi and Noctor (2016) on *Arabidopsis*, suggest a tradeoff in plant resistance to pathogens and herbivorous insects, whereby plants become more tolerant to pathogens but also more vulnerable to herbivores with chewing mouthparts.

Growth under elevated CO_2 alters the foliar profiles of secondary metabolites relative to growth under ambient CO_2 (Lindroth 2012). Discerning generalization about the direction of response and classes of compounds affecting such empirical observations, however, is difficult because the chemical defenses produced by plants are idiosyncratic, varying widely from species to species (Berenbaum 1995). Moreover, many such defenses become apparent only upon induction by an insect herbivore; while constitutive expression of secondary metabolites varies by species and CO_2 environment, the element of defense-on-demand, or dynamic induction, adds another dimension of complexity (Meldau et al. 2012). Rather than investigating patterns among chemical phenotypes, a more practical approach to understanding how elevated CO_2 modulates defense may be to consider the physiological mechanisms that underlie plant perception and response to stress. Because the signaling pathways responding to herbivory are more similar among plants than the wide array of chemical defenses produced, this strategy, implicit in many recent reviews on plant secondary metabolism (e.g., Meldau et al. 2012; Zavala et al. 2013; Schuman and Baldwin 2016), may more readily reveal generalizable responses and underlying mechanisms.

Some generalities in the induction of defense become apparent at the level of plant hormones, although these systems remain quite complex. While plants produce a vast array of secondary metabolites, the induction of chemical defense rests primarily on the activity of two major and highly conserved defense hormones, jasmonic acid and salicylic acid (Thaler et al. 2012). Jasmonic acid and salicylic acid interact with one another as well as with the gamut of other major plant hormones, in particular, gibberellic acid (GA), cytokinins, abscisic acid (ABA), and ethylene (Erb et al. 2012). Moreover, both jasmonic acid and salicylic acid serve functions beyond defense; jasmonic acid is integrated with plant growth while salicylic acid is a determining factor in flowering time (Wasternack and Hause 2013; Vicente and Plascenia 2011). Recently, Zhang et al. (2015) suggest that cross-talk between defense hormones accounts for observed differences in plant resistance to pathogens and insect herbivores under variable CO_2. Thus, though complex, the existence of a regulatory link between the growth and defensive roles of primary and secondary metabolism can be discerned in the concerted activity of plant hormones.

Increasing CO_2 concentration affects both the growth and defense habits of C3 plants, but the regulatory connection between the two roles is not well understood (Zavala et al. 2013). Plants grown under elevated CO_2 exhibit increased foliar concentrations of salicylic acid (Casteel et al. 2012; Zhang et al. 2015; Mhamdi and Noctor 2016). This is thought to suppress the activity jasmonic acid and the expression of its associated chemical defenses (Casteel et al. 2012; Zhang et al. 2015). In turn, the induction of salicylic acid is known to rest on the redox environment within plant cells (Leon et al. 1995; Mateo et al. 2006). Under perception of attack from a biotic agent, the stimulus for a defense response typically begins with a burst of reactive oxygen species (ROS) (Lamb and Dixon 1997). Such 'pulses' of ROS in plant cells can trigger phosphorylation cascades conducted by mitogen activated protein kinases (MAPKs) (Jonak et al. 2002; Apel and Hirt 2004). While the role of CO_2 in affecting either the activity of ROS or mapk's is not well resolved, both photosynthesis and photorespiration are two major sources of CO_2-dependent ROS production in plant cells, and numerous reviews suggest links between ROS production in plant cells and hormonal regulation (e.g., Kerchev et al. 2012; Foyer and Noctor 2009). That exposure to elevated CO_2 affects photosynthesis and photorespiration and consequently ROS produced by these pathways, and that ROS affects defense related hormonal signaling provides a hypothetical link to explain the co-regulation of primary and secondary metabolism.

Although how elevated CO_2 modulates the defense hormone salicylic acid is unknown, decades of research in photosynthesis, signal transduction and physiological ecology have produced a body of literature that imply a connection. The first objective of this chapter is to review the literature that implicates a causal relationship between atmospheric chemistry and plant defense habit. Such a review necessitates emphasis on identifying hypothetical connections that are plausible and informative, but experimentally undefined. Based on a synthesis of available literature, we present the novel hypothesis that: (1) Elevated CO_2 in combination with variable light causes transient excess energy in electron transport to drive increased production of hydrogen peroxide; and, (2) This production of hydrogen peroxide acts as a molecular signal that is transduced through MAPKs to stimulate biosynthesis of salicylic acid. A challenge to testing this hypothesis is that it depends on difficult and highly uncertain measurements of excess energy flow in the light reactions of photosynthesis and intra-cellular concentrations of hydrogen peroxide. Recognizing this methodological challenge, a second objective of this chapter is to propose experimentally falsifiable hypotheses, designed to link previously existing knowledge on plant primary and secondary metabolism. Although this chapter is not intended as a review of experimental methods and protocols, some discussion and explanation of laboratory methods is necessary to describe practically feasible experimental approaches to testing the hypothesis.

How does Elevated CO_2 Affect Primary Metabolism: Setting the Stage

The enzyme responsible for assimilating both CO_2 and O_2 into plant metabolism, Ribulose-1,5-bisphosphate-Carboxylase/Oxygenase (RuBisCO) is thought to be the most abundant protein on Earth (Ellis 1979; Raven 2013). As the predominant entry point for carbon into the biosphere, the active site of RuBisCo represents a transition between atmospheric chemistry and the Earth's ecosystem. In C3 plants, the oxygenation reaction of RuBisCO competes with its crucial carboxylation reaction (Laing 1974). The carboxylation and oxygenation pathways differ in the immediate metabolic fate of their reactants; one product of fixation enters the Calvin-Benson-Basham (CBB) Cycle and the other enters photorespiration, respectively (Weissbach 1956; Ogren and Bowes 1971). While the Calvin Cycle takes place exclusively within the chloroplast, metabolites in the photorespiratory pathway progress through several organelles; from chloroplast to peroxisome, then mitochondria and finally returning to the chloroplast (Fig. 1).

The absence of photorespiration as a 'sink' for energy acquired from light could indirectly favor the reduction of oxygen in the chloroplast. The

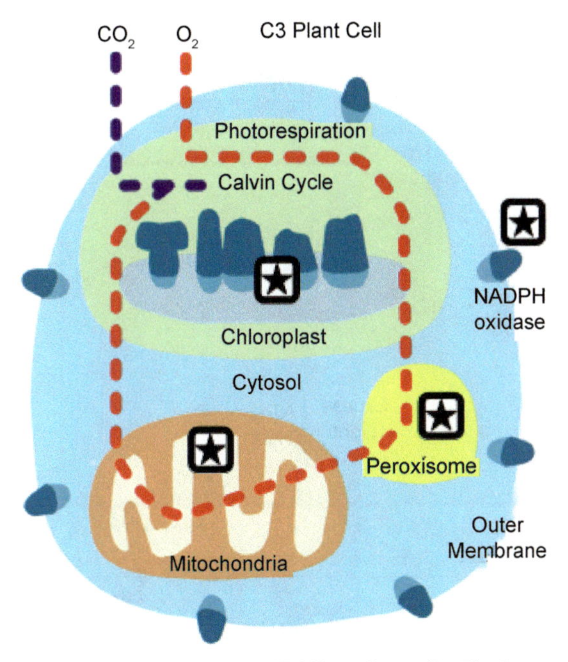

Figure 1. Sources of reaction oxygen species (ROS) in plant cells. The four major sites for the production of ROS are labelled with a star. Metabolic routes followed by the Calvin Cycle and photorespiration are represented as dashed lines.

metabolic pathways that incorporate atmospheric CO_2 and O_2 into plant metabolism differ in their energetic requirements (Farquhar et al. 1980). Per molecule of CO_2 or O_2, respectively, removed from the atmosphere, the Calvin Cycle consumes slightly less ATP and slightly more NADPH relative to photorespiration. Under non-photorespiratory conditions, the ratio of ATP to NADPH consumption by photosynthesis is approximately 21:14, which increases to approximately 21:13 under photorespiratory conditions (Foyer et al. 2012), a small but potentially important change. This difference in stoichiometry means that as an increase in the supply of CO_2 progressively stimulates carboxylation and reduces O_2 consumption by photorespiration, the overall ratio of ATP to NADPH required for both processes declines. Because ATP production, non-photochemical quenching (NPQ), and cyclic electron flow all form a regulatory circuit with one another (Fig. 2; Foyer et al. 2012; Niyogi 1999; Kanazawa and Kramer 2002), one possible consequence of this change in stoichiometry is the exhaust of electrons onto oxygen to form superoxide in the chloroplast stroma (Asada 1999).

To adjust for changes in ATP demand, chloroplasts modulate cyclic electron flow by routing electrons from photosystem I (PSI) through the cytochrome b6f complex (Munekage et al. 2004; Miyake et al. 2005;

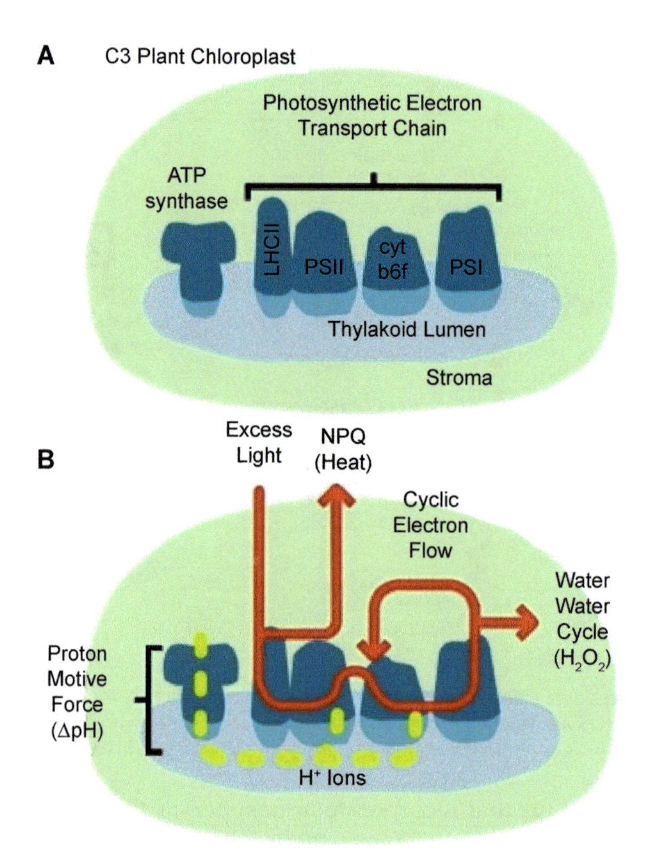

Figure 2A: Schematic of major components of the photosynthetic electron transport chain. 2B: The relationship between electron-flow, proton-motive force and photoprotection (NPQ) in C3 plant chloroplasts. The red line represents flow of energy, either as heat or excitation transfer, while the yellow dashes represent hydrogen ions generated by energy flow through the PETC.

Walker et al. 2014). Electron flow through the cytochrome b6f complex pumps hydrogen ions into the thylakoid lumen, thus generating the proton motive force (pmf) to drive ATP synthase. Although Miyake et al. (2005) and Walker et al. (2013) report that cyclic electron flow increases as pressure of CO_2 declines, to our knowledge no study has addressed the question of how increasing CO_2 affects cyclic electron flow (but see Kanazawa and Kramer 2002). One might suppose that as atmospheric CO_2 increases, the metabolic demand for ATP declines, reducing cyclic electron flow, with the effect that the lumen does not acidify as quickly as it would under conditions of low CO_2.

By lowering pmf, elevated CO_2 relaxes non-photochemical quenching (NPQ) of excess excitation energy in the photosynthetic electron transport chain (PETC) (Miyake et al. 2005; Kanazawa and Kramer 2002). When plants are exposed to high light, or sudden changes in light intensity, the increased flow of electrons through the PETC rapidly acidifies the lumen beyond the proton motive force required for ATP synthesis (Ort 2001). When the lumen acidifies in this manner, NPQ is induced by the rapid acidification of the thylakoid lumen; the photosynthetic light harvesting complexes (LHCs) change conformation to favor dissipation of excitation energy as heat through xanthophyll pigments in associated light harvesting complexes (Niyogi 1999; Li et al. 2002). The relaxation of NPQ with increasing CO_2 would be the proximate source of additional excess excitation energy in the PETC, under the presumption that the plant loses some efficiency of heat dissipation as an alternative sink for excess energy.

In the absence of a means to dissipate excess excitation energy, it is conceivable that the chloroplast would divert electrons directly to O_2 creating reactive oxygen species (ROS), thereby altering the redox environment within the chloroplast. The Water–Water cycle, also known as the Mehler reaction or pseudocyclic electron flow, describes the transfer of electrons from the PETC to oxygen to form superoxide, followed by antioxidant quenching by ascorbate to glutathione (Asada 1999). Since its core reaction was first identified in the 1950's as the Hill reaction (Hill 1950) the extent of activity and the physiological role played by the Mehler reaction (Mehler 1951) has been a source of debate (for reviews, see: Niyogi 1999; Foyer and Noctor 2012; Heber 2002). To date, there is little experimental evidence that the Water–Water reaction dissipates excess energy, at least not in a manner that could be called a biochemical 'sink' for energy (Ruuska et al. 2000; Driever and Baker 2011). For instance, Heber (2002) argues that the Water–Water cycle prevents transport components of the PETC from becoming over-reduced, particularly under conditions of excess energy, such that cyclic electron flow can properly operate. From this perspective, the Water–Water cycle is perhaps best understood as an exhaust conduit for excess electrons in the PETC. Hence, rather than a true buffer to dissipate excess energy from light, variation in the activity of the Water–Water cycle could be meaningful to intercellular signaling as a source of reactive oxygen species (ROS) in plant cells.

The Linkage between ROS and Defense

Since the late 1990's, reactive oxygen species (ROS) have received field-wide (e.g., Apel and Hirt 2004; Mittler et al. 2011; Foyer and Noctor 2005) and sometimes divisive (see Alpi et al. 2007 vs. Trewavas 2007) discussion as the proximal basis for perception of—and responses to—the surroundings of individual plants. At heart, the redox-based signal transduction consists of the discharge of electrons through a biological substrate; tasked with quenching electrochemical energy, plant cells pass electrons through complex sequences of reduction and oxidation reactions (Foyer and Noctor 2005). The intracellular sequence of reactions giving rise to ROS spans chloroplast, peroxisome, mitochondria, outer membrane and nucleus (Fig. 1; Cheeseman 2007). By maintaining overall redox homeostasis, plants can sense environmental disturbance as fluctuations in redox state within individual cells (Foyer and Noctor 2005).

Photosynthesis is the first and second largest sources of hydrogen peroxide in plant cells: photorespiration and the Water–Water cycle, respectively (Fig. 1., Foyer and Noctor 2003). During photorespiration, 2-PGA formed in the chloroplast is transported to the peroxisome where glycolate is oxidized to form glyoxalate and hydrogen peroxide (Ogren 1984). The Water–Water cycle describes the production of ROS as a function of electron flow through the PETC and subsequent quenching of hydrogen perxoide by ascorbate within the chloroplast (Asada 1999). Because the Water–Water cycle modulates ROS production in the chloroplast, where the cascade of events leading the production of SA is located, this cycle may play a key role communicating changes in the environment affecting primary metabolism to changes in secondary metabolism that affect susceptibility to herbivory. Beyond photosynthesis, ROS is produced by electron-transport in mitochondrial respiration (Møller 2001), albeit in much lower volume than from photosynthesis and photorespiration (Foyer and Noctor 2003). Unlike photorespiration, photosynthesis and mitochondrial respiration, NADPH-oxidases are activated under specific circumstances. Bound to the outer membrane of plant cells, NADPH-oxidases mediate many growth functions as well as the first lines of defense during invasion from biotic agents (Marino et al. 2012).

Biotic stress universally downregulates photosynthesis genes (Bilgin et al. 2010), but it is unclear whether this regulatory relationship results from retrograde signaling, anterograde signaling, or some combination of the two. Chloroplasts give rise to a variety of signals that influence expression of genes and hormone synthesis in the nuclei of plant cells (Demmig-Adams et al. 2014; Tikkanen et al. 2014; Gollan et al. 2015). Such signaling is considered 'retrograde' because the chloroplast generates the impetus for nuclear gene expression, whereas 'anterograde' signaling occurs when the nucleus manipulates processes occurring in the chloroplast (Woodson and Chory 2008).

The conventional model of nuclear and chloroplast gene expression following recognition of a biotic agent generally assumes that anterograde signaling is responsible for plant responses to stress. Biotic stress—from pathogens or insect herbivores—generates ROS (Torres et al. 2006; Bi and Felton 1995). When plant cells recognize attack through microbial activated molecular patterns (MAMPs), they often trigger adjacent outer-membrane-bound NADPH-oxidases, which release superoxide onto the cellulose fibers comprising the plant cell wall (Marino et al. 2012). This burst of ROS is recognized as signal event in its own right, because it stimulates salicylic acid biosynthesis, a major plant defense hormone (Lamb and Dixon 1997). Such a signal cascade originating in the outer cell membrane and traveling through the nucleus to regulate chloroplast function would represent an example of anterograde signaling.

In contrast to anterograde signaling, an environmental stimulus originating in the chloroplast and traveling to the nucleus would represent retrograde a signal. An imbalance in the energy distribution in PETC disrupts redox homeostasis, producing ROS, which then influence nuclear gene expression (Gollan et al. 2015). Demmig-Adams et al. (2014) postulate that photoprotective mechanisms modulate this production of ROS and therefore play an important role in physiological regulation, beyond their immediate function in dissipating excess excitation energy. The chloroplast is directly involved in the biosynthesis pathways of two major defense hormones, salicylic and jasmonic acid (Seyfferth and Tsuda 2014; Wasternack and Hause 2013). Because both retrograde and anterograde signal transduction is mediated through the oxidative environment of the plant cell, it is possible that the two types of signals would interact with one another during the induction of defense.

Salicylic acid is an essential mediator between gene transcription and the redox environment within plant tissue (i.e., Leon et al. 1995; Mateo et al. 2006; Tada et al. 2008). NPR1, a master regulating protein responsible for initiating transcription of defense genes, is activated by changes in the redox state of plant cells (Tada et al. 2008). In turn, concentration-dependent reception of salicylic acid by NPR3 and NPR4 determines which defense genes NPR1 activates (Fu et al. 2012). Expression of enzymes driving phenylpropanoid metabolism, in particular, are regulated by salicylic acid (Dixon et al. 2002). During the hypersensitive response (HR) of plant cells, hydrogen peroxide rapidly accumulates in foliar tissue (Levine et al. 1994). When salicylic acid crosses a concentration threshold, the NPR complex activates genetic expression of HR and systemic acquired resistance (SAR) responses (Fu et al. 2012).

Salicylic acid, itself a simple phenolic compound, directs phenylalanine lyase (PAL) metabolism to generate a multitude of phenolic compounds (Dixon et al. 2002). The 6-carbon ring structures common to all such phenolic compounds are characterized by high chemical stability; this core attribute of chemical stability fills a wide variety of advantageous biological functions. Phenolic compounds increase antioxidant capacity of plants because they can accept electrons without losing structural integrity (Dixon and Palva 1995). Similarly, the class of flavonoid compounds absorb UV-radiation, thereby acting as a sunscreen (Dixon and Palva 1995). Lignin, a major component of woody tissue, is a 'super-molecule' created by linking simple phenylpropanoid compounds in a chaotic repeating fashion (Boerjan et al. 2003). As a defense against herbivorous insects, many plants produce polyphenol-oxidases (PPOs), enzymes which form cross-links between phenolic compounds (Appel 1993). When ingested, PPOs form difficult-to-digest masses of phenolic compounds inside insect guts thereby impeding rate of herbivory (Appel 1993).

When plants perceive stress from pathogens or insects, their chemical defense response frequently rests upon a highly-conserved tradeoff between defense hormones salicylic acid and jasmonic acid (Thaler et al. 2012). Herbivores with chewing mouthparts can possess mechanisms to induce salicylic acid in their host-plants, so as to bypass defenses based on jasmonic acid signaling (Musser et al. 2002; Kästner et al. 2014). Conversely, *Pseudomonas syringae* is a microbial pathogen that inoculates its hosts with coronatine, a chemical mimic of jasmonic acid, thereby suppressing induction of salicylic acid (Zheng et al. 2012). Some studies document that plants grown under elevated CO_2 are more resistant against infection from viral pathogens as a consequence of defenses upregulated by salicylic acid (Matros et al. 2006; Huang et al. 2012). Zhang et al. (2015), find that elevated CO_2 suppresses jasmonic-acid based defense in tomato by upregulating salicylic acid. Moreover, Mhamdi and Noctor (2016) demonstrate that elevated CO_2 upregulates salicylic acid through redox-linked pathways (Mhamdi and Noctor 2016).

Soybean plants grown under elevated CO_2 at the soyFACE field experiment exhibit changes that match characteristics of altered ROS signaling from the chloroplast. Although not isolated to the chloroplast, Cheeseman (2006) found that leaf tissue grown under elevated CO_2 contains more hydrogen peroxide than leaf tissue grown under ambient CO_2, indicating a change in redox state. Qiu et al. (2008) report that protein isolated from leaves of the same experimental treatment exhibit increased carbonylation, a symptom of increased carbonic acid content in leaf tissue that could be caused by greater concentration of hydrogen peroxide. Similarly, foliar concentration salicylic acid is known to track the redox state of plant tissue, and Casteel et al. (2012) observe that foliar concentration of salicylic acid in soybean is higher under elevated CO_2 relative to control. Hence, it is possible that CO_2 modulates the redox environment within soybean tissue and thus influence hormonal regulation of plant chemical defense.

How does the Production of Photochemical ROS Modulate Defense Hormones?

Reactive oxygen species (ROS) signaling controls different biological processes such as responses to biotic and/or abiotic stimuli (Mittler et al.

2011). Although ROS can be accumulated in chloroplast and mitochondria of plants (Apel and Hirt 2004; Mittler et al. 2004), these reactive species are mainly produced by cell wall NADPH oxidases and peroxidases (Apel and Hirt 2004; Nurnberger et al. 2004). Because of the potential toxicity of ROS, non-toxic levels of these species must be maintained in a delicate balancing between ROS production and the metabolic counter-process involving ROS-scavenging enzymes (Mittler et al. 2004). Since plant responses might be regulated by temporal and spatial coordination between ROS and other signals, ROS function as secondary messengers that induce important signaling pathways.

In response to ROS accumulation, early signaling events in plants include increased flux of Ca^{2+} into the cytosol, activation of mitogen-activated protein kinases (MAPKs), and protein phosphorylation (Benschop et al. 2007). MAPKs play major role in signal transduction of diverse stress responses. Three consecutive elements (MAPKKK, MAPKK or MEK and MAPK) compose the signal cascade in which MAPK is finally phosphorylated and activated (Hamel et al. 2006). While the MAPKK4-MAPK3/6 module is known to play role in ROS production by acting upstream of NADPH oxidase, accumulation of H_2O_2 activates MAPK3 and MAPK6 (Kovtun et al. 2000). Although MAPKs are known to be activated by ROS molecules, the mechanism behind the specific activation of MAPK cascade by ROS is not clear (Jalmi and Sinha 2015).

The interactions between ROS and MAPKs have been observed in different plant species. H_2O_2 activates MAPK3 and MAPK6 in rice and gets activated by upstream kinase MAKK6. Pathogen attack induced the accumulation of ROS and activated *Arabidopsis* MAPK3, MAPK4, and MAPK6. Pathogen attack in *Arabidopsis* produced ROS and activated MAPK through the cascade of MEKK1-MEKK4/5-MAPK3/6 (Asai et al. 2002). This cascade induced resistance to a fungal pathogen as well as increasing tolerance to abiotic stresses (Kumar and Sinha 2013; Sheikh et al. 2013; Singh and Jwa 2013). MAPK ultimately phosphorylate and activate several downstream targets like transcription factor, other kinases, phosphatases, and cytoskeleton associated proteins (Hamel et al. 2006; Rodriguez et al. 2010).

MAPK activation and the ROS burst are two early events that trigger plant immunity. Pathogen and herbivore damage are perceived by leaves and lead to rapid transcription and activation of MAPK signaling pathways that induce JA/ET- and SA-regulated defenses in plants (Kandoth et al. 2007; Wu et al. 2007; Liu et al. 2014; Petersen et al. 2000; Liu et al. 2011). Whereas both MAPK3 and MAKP6 redundantly control SA and JA in different plant species, only MAPK6 regulates ET emission (Wu et al. 2007; Kandoth et al. 2007). Silencing MAPK6 but not MAPK3 diminished herbivory-induced ET levels in *N. attenuata* and *Arabidopsis* by decreasing other kinases expression, like CDPKs (Wu et al. 2007; Schafer et al. 2011). In field-grown soybean, after few minutes of stink bug damage to developing seeds, MAPK6 phosphorylation was enhanced in addition to induction of JA/ET and SA (Giacometti et al. 2016). Although MAPK3 and MAPK6 are involved in the amplification of defensive reactions, MAPK4 takes part in repressing SA biosynthesis. MAPK4 in *Arabidopsis* deters SA signaling pathway by inhibiting EDS1 and PAD4 proteins and liberating the JA/ET pathway (Petersen et al. 2000; Brodersen et al. 2006). GmMAPK4-silenced soybean accumulated SA and SA-regulated genes were up-regulated in leaves (Liu et al. 2011).

A New Hypothesis Linking Primary and Secondary Metabolism

Imbalances Lead to ROS in the Chloroplast Through Water–Water Cycle

Per molecule removed from the atmosphere, CO_2 fixation consumes proportionally less ATP than O_2 fixation (Foyer et al. 2009; Foyer et al. 2012). To adjust for changes in energetic demand caused by variation in gas composition, C3 plants decrease cyclic electron flow with rising CO_2 concentration (Walker et al. 2015; Miyake et al. 2005). Reducing cyclic electron flow relaxes the proton motive force (pmf or ΔpH) across the thylakoid membrane, thereby slowing ATP production in the chloroplast. This relaxation in pmf is known to suppress NPQ mechanisms (Munekage et al. 2004; Miyake et al. 2005), although the effect has not been tested in context of rising atmospheric CO_2. A reduced capacity for NPQ could mean that plants lose heat-dissipation as a means for disposing of excess excitation energy; the remaining question, then, is whether this loss would increase allocation of excess energy to the Water–Water cycle.

Imbalances are Exacerbated in Dynamic Light Environments

Asada (1999) defines 'excess energy' in PETC as any electron flow that cannot be allocated to the combined sinks the Calvin Cycle and photorespiration. Under steady light conditions, plants balance energy intake against the energy demanded by fixation of atmospheric carbon and oxygen. However, during sudden transitions from dark to bright light conditions, such as would occur during cloud-breaks or sunflecks due to moving canopy cover, the sudden influx of light energy into the PETC could outpace demand from its primary sinks (Leakey et al. 2003; Watling et al. 1997). During such times, NPQ plays an important role in dissipating excess excitation energy as heat from the light-harvesting complexes of the PETC (Watling et al. 1997).

During the Mehler reaction, or Water–Water cycle, excess excitation energy in the photosynthetic electron transport chain (PETC) is transferred to oxygen and then to water to form hydrogen peroxide (Asada 1999). The difference between measured electron transport rates (J) and theoretical rates calculated from gas exchange (J_c or J_g) has been attributed to the Mehler reaction (Krall and Edwards 1991; Ruuska et al. 2000). Thus, if the difference between J and J_g increases when plants are exposed to sudden light variation, the implication is that plants absorb more excitation energy than can be immediately directed to the combined sink of photosynthesis and photorespiration. By the same token, when the PETC departs from redox homeostasis- or non-steady-state photosynthesis—the efficiency of light harvest (Φ_{PSII}) should have a greater rate of change than the efficiency of carbon assimilation (Φ_{CO2}).

Perturbation to redox-homeostasis, or steady-state photosynthesis, is thought to form the basis for retrograde signaling by the chloroplast (for review see Gollan et al. 2015). During chloroplast retrograde signaling, environmental signals originate in the PETC and are transduced through various redox-dependent pathways to ultimately control nuclear gene expression. If the initial signal stimulus in the PETC is modulated by CO_2 concentration, what follows is that retrograde signaling from the chloroplast in response to environmental stress would differ as well. Salicylic acid, in particular, has a well-documented relationship with redox state of plant tissue and expression of genes related to innate immunity (respectively: Mateo et al. 2006; Leon et al. 1995; Vlot et al. 2009).

Are there other Environmental Factors that Drive the Imbalance?

Although this hypothesis focuses on acquisition of light energy, in principle any environmental factor that upsets the balance between energy supply and demand in the PETC would result in an ROS burst from the chloroplast. Under suboptimal temperature conditions, for instance, Fryer et al. (1998) observe that maize plants exhibit a nonlinear relationship between Φ_{PSII} and Φ_{CO2}. Situations that cause rapid closure of stomata could conceivably alter the ratio of CO_2 and O_2 in stomatal cavities, thus restricting CO_2 supply to adjoining cells. The effects of stomata closure during stress deserve careful consideration, as Farquhar and Sharkey (1982) indicate that stress-induced decreases in stomatal conductance rarely influence rates of carbon assimilation. Under dynamic or temporary conditions of stomata closure, however, a rapid shift in intercellular pressure of CO_2 could have the effect of generating differential ROS from chloroplast and peroxisome, an event outlined in more detail by Kangasjärvi et al. (2012).

Considering that some studies observe an influence of insect herbivory on stomatal conductance (e.g., Nabity et al. 2013; Meza-Canales et al. 2017), the role of stomatal closure during stress on redox-based signaling remains an open question.

Why should Photosynthesis and Defense be Linked? Plant Carbon/Energy Allocation?

The possibility that chloroplast retrograde signaling modifies plant defense responses appeals to the core tenets of optimal defense theories, as described by Zangerl and Bazzaz (1992). Since the 1970's, numerous theories have been proposed to explain how plants allocate energy and material resources between the primary functions of growth and secondary functions of defense (Schuman and Baldwin 2015). The common denominator among such theories holds that plant defense exacts some cost on plant growth, leading to a tradeoff between growth and defense; plants must somehow balance the two functions in order to optimize their evolutionary fitness. Thus, because photosynthesis drives plant growth, the idea that regulation of defense would occur as a function of perturbation to photosynthesis matches ongoing research on the connection between primary and secondary metabolism in plants.

Experimental Approaches to Testing the Hypothesis

The hypothesis that C3 plant chloroplasts could modulate plant secondary metabolism in a CO_2-dependent manner encompasses both proximate photosynthetic mechanisms and ultimate effects on interaction between plants and insects. The two proximate aspects of this hypothesis are that: (1) Elevated CO_2 in combination with variable light causes transient excess energy in electron transport to drive increased production of hydrogen peroxide; and that (2) This increase in production of hydrogen peroxide acts as a molecular signal that is transduced through MAPKs to stimulate biosynthesis of salicylic acid. Hence, *if* increasing CO_2 amplifies the production of ROS by photosynthesis and *if* such an increase in ROS acts as a signal that is meaningful to the biosynthesis of defense hormones, *then* one might predict that the ultimate expression of plant chemical defense against herbivorous insects would likewise depend on concentration of CO_2 in the atmosphere. Hence, tests of this hypothesis are predicated upon measuring interactions between the light and CO_2 environment of plants and their effects on the production of ROS, MAPKs, defense hormones, production of secondary metabolites and plant-insect interactions.

A major obstacle to testing the hypothesis that hydrogen peroxide produced in chloroplasts ultimately influences plant-insect interactions is the chemical instability of reactive oxygen species: They are very difficult, if not impossible, to isolate and quantify. Both Cheeseman (2006) and Noctor et al. (2016) have published reviews on the topic of measuring ROS in plant tissue, and have concluded that current methods are too error-prone to be trustworthy. Hence, even if one were to use described methods for chemically separating and quantifying ROS in plant tissue, the results would likely meet with implicit distrust from the scientific community. A more effective approach than measuring ROS directly may be to focus on documenting variation in the processes that give rise to ROS, or to seek the immediate effects of the production. Mhamdi and Noctor (2016), for example, use such an indirect measurement strategy by documenting endogenous pools of the antioxidants glutathione and ascorbic acid to demonstrate that the induction of salicylic acid by elevated CO_2 in *Arabidopsis* is redox-dependent.

At the level of photosynthesis, integrated measurements of gas exchange and chlorophyll fluorescence can determine whether elevated CO_2 increases allocation of electrons to oxygen in dynamic light environments. If cyclic electron flow around PSI declines as CO_2 concentrations increase, then induction of NPQ in C3 plants should be relaxed under elevated CO_2 relative to ambient CO_2. Moreover, if the activity of the Water–Water cycle during sudden transitions from steady to non-steady state photosynthesis depends on CO_2 level, then for C3 plants the difference between electron transport rate (J) and theoretical electron transport rate (J_g) should be higher under elevated CO_2 relative to ambient CO_2. In other words, Φ_{PSII} should outpace Φ_{CO2} when plants are exposed to sudden increases in light intensity, especially so with increasing CO_2. The theoretical rate of electron transport (J_g) can be calculated based on a derivation of the Farquhar-von Caemmerer-Berry model of photosynthesis, as in Ruuska et al. (2000). Similarly, when Φ_{PSII} displays a non-linear relationship with Φ_{CO2}, the implication is that the PETC is acquiring more energy than can be immediately directed towards carbon metabolism, as in Fryer et al. (1998).

At the level of intra-cellular signaling, imaging techniques can be used to visualize ROS production in leaf tissue while chromatographic methods can quantify concentrations of major defense hormones. To visualize ROS, diamine-benzidine (DAB) stain reacts with H_2O_2 in the presence of peroxidases to form a brown precipitate. If elevated CO_2 causes differential production of ROS, then leaf tissue grown exposed to sudden dark-to-light transition should appear dark when treated with DAB solution, relative to untreated leaves or leaves treated under ambient CO_2. Similarly, if the induction of salicylic acid by fluctuation in light levels depends on CO_2, then the magnitude of induction of salicylic in C3 plants in dynamic light environments should be increased under elevated

CO_2 relative to ambient CO_2. At the same time, induction of jasmonic acid should be relatively suppressed.

At the level of plant-insect interactions, chemical analysis of secondary metabolites and bioassays with insect herbivores can determine the palatability of plant tissue changes depending on light and CO_2 environment. For instance, if dynamic light and elevated CO_2 interact so as to compromise plant chemical defenses, then insect herbivores would be expected to both prefer and consume more of plant tissue grown under dynamic light and elevated CO_2 than plant tissue grown under steady light and ambient CO_2. Likewise, the profiles of secondary metabolites exhibited by plants grown under differential light and CO_2 regimes should differ from one another.

At all stages of hypothesis testing, genetic manipulation of plants may help to resolve the signaling pathways transduced from initial perturbations of photosynthesis to ultimate effects on plant-insect interactions. To name just a few possibilities, the *Arabidopsis* biological resource center (ABRC) at www.abrc.osu.edu maintains a stock of genetically altered *Arabidopsis* lines that pertain to the hypothesis discussed in this chapter. Among those lines of *Arabidopsis* are NPQ mutants which are deficient in the induction of nonphotochemical quenching (originally developed by Niyogi et al. 1998) and ascorbate mutants which are deficient in production of the antioxidant ascorbate (originally developed by Conklin et al. 2000). By relaxing the induction of NPQ through genetic manipulation, one might expect to replicate the effects of elevated CO_2 on induction of foliar salicylic acid by dynamic light. Similarly, by reducing available pools of ascorbic acid, one might expect to amplify hydrogen peroxide in leaf tissue and thus increase foliar concentration of salicylic acid. Given the broad nature of this hypothesis, many more genetic targets may be possible; the suggestions provided here are intended to serve as examples of how genetic manipulation in plants might be applied to help in testing.

References

Alpi, A., N. Amrhein, A. Bertl et al. 2007. Plant neurobiology: no brain, no gain? Trends Plant Sci. 12(4): 135–136.

Apel, K. and H. Hirt. 2004. Reactive oxygen species: metabolism, oxidative stress, and signal transduction. Annu Rev Plant Biol. 55: 373–399.

Appel, H.M. 1993. Phenolics in ecological interactions: the importance of oxidation. J Chem Ecol. 19(7): 1521–1552.

Asada, K. 1999. The water-water cycle in chloroplasts: scavenging of active oxygens and dissipation of excess photons. Annu Rev Plant Biol. 50(1): 601–639.

Asai, T., G. Tena, J. Plotnikova et al. 2002. MAP kinase signaling cascade in Arabidopsis innate immunity. Nature. 415: 977–983.

Berenbaum, M.R. 1995. The chemistry of defense: theory and practice. Proc Natl Acad Sci USA. 92(1): 2–8.

Bi, J.L. and G.W. Felton. 1995. Foliar oxidative stress and insect herbivory: primary compounds, secondary metabolites, and reactive oxygen species as components of induced resistance. J Chem Ecol. 21(10): 1511–1530.
Bilgin, D.D., J.A. Zavala, J.I.N. Zhu et al. 2010. Biotic stress globally downregulates photosynthesis genes. Plant Cell Environ. 33(10): 1597–1613.
Boerjan, W., J. Ralph and M. Baucher. 2003. Lignin biosynthesis. Ann Rev Plant Biol. 54(1): 519–546.
Casteel, C.L., L.M. Segal, O.K. Niziolek et al. 2012. Elevated carbon dioxide increases salicylic acid in Glycine max. Environ Entomol. 41(6): 1435–1442.
Cheeseman, J.M. 2006. Hydrogen peroxide concentrations in leaves under natural conditions. J Exp Biol. 57(10): 2435–2444.
Cheeseman, J.M. 2007. Hydrogen peroxide and plant stress: a challenging relationship. Plant Stress. 1(1): 4–15.
Conklin, P.L., S.A. Saracco, S.R. Norris et al. 2000. Identification of ascorbic acid-deficient Arabidopsis thaliana mutants. Genetics. 154(2): 847–856.
Demmig-Adams, B., J.J. Stewart and W.W. Adams. 2014. Multiple feedbacks between chloroplast and whole plant in the context of plant adaptation and acclimation to the environment. Philos T Roy B. 369(1640): 20130244.
Dermody, O., B.F. O'Neill, A.R. Zangerl et al. 2008. Effects of elevated CO_2 and O_3 on leaf damage and insect abundance in a soybean agroecosystem. Arthropod Plant Interact. 2: 1125–1135.
Dixon, R.A. and N.L. Paiva. 1995. Stress-induced phenylpropanoid metabolism. Plant Cell. 7(7): 1085.
Dixon, R.A., L. Achnine, P. Kota et al. 2002. The phenylpropanoid pathway and plant defence—a genomics perspective. Mol Plant Pathol. 3(5): 371–390.
Ellis, R.J. 1979. The most abundant protein in the world. Trends Biochem Sci. 4(11): 241–244.
Erb, M., S. Meldau and G.A. Howe. 2012. Role of phytohormones in insect-specific plant reactions. Trends Plant Sci. 17(5): 250–259.
Farquhar, G.D., S. von Caemmerer and J.A. Berry. 1980. A biochemical model of photosynthetic CO_2 assimilation in leaves of C3 species. Planta. 149.1: 78–90.
Foyer, C.H. and G. Noctor. 2003. Redox sensing and signaling associated with reactive oxygen in chloroplasts, peroxisomes and mitochondria. Physiol Plantarum. 119(3): 355–364.
Foyer, C.H. and G. Noctor. 2009. Redox regulation in photosynthetic organisms: signaling, acclimation, and practical implications. Antioxid Redox Sign. 11(4): 861–905.
Foyer, C.H., J. Neukermans, G. Queval et al. 2012. Photosynthetic control of electron transport and the regulation of gene expression. J Exp Bot. 63(4): 1637–1661.
Fryer, M.J., J.R. Andrews, K. Oxborough et al. 1998. Relationship between CO_2 assimilation, photosynthetic electron transport, and active O_2 metabolism in leaves of maize in the field during periods of low temperature. Plant Physiol. 116(2): 571–580.
Fu, Z.Q., S. Yan, A. Saleh et al. 2012. NPR3 and NPR4 are receptors for the immune signal salicylic acid in plants. Nature. 486(7402): 228–232.
Gollan, P.J., M. Tikkanen and E.M. Aro. 2015. Photosynthetic light reactions: integral to chloroplast retrograde signaling. Curr Opin Plant Biol. 27: 180–191.
Guo, H., Y. Sun, Q. Ren et al. 2012. Elevated CO_2 reduces the resistance and tolerance of tomato plants to *Helicoverpa armigera* by suppressing the JA signaling pathway. PLoS One. 7(7): e41426.
Hamel, L.P., M.C. Nicole, S. Sritubtim et al. 2006. Ancient signals: comparative genomics of plant MAPK and MAPKK gene families. Trends Plant Sci. 11: 192–198.
Hamilton, J.G., O. Dermody, M. Aldea et al. 2005. Anthropogenic changes in tropospheric composition increases susceptibility of soybean to insect herbivory. Environ Entomol. 34: 479–485.
Heber, U. 2002. Irrungen, Wirrungen? The Mehler reaction in relation to cyclic electron transport in C3 plants. Photosynth Res. (1-3): 223–231.

Huang, L., Q. Ren, Y. Sun et al. 2012. Lower incidence and severity of tomato virus in elevated CO_2 is accompanied by modulated plant induced defence in tomato. Plant Biol. 14(6): 905–913.
Kanazawa, A. and D.M. Kramer. 2002. *In vivo* modulation of nonphotochemical exciton quenching (NPQ) by regulation of the chloroplast ATP synthase. Proc Natl Acad Sci USA. 99(20): 12789–12794.
Jalmi, S.J. and A.K. Sinha. 2015. ROS mediated MAPK signaling in abiotic and biotic-striking similarities and differences. Front Plant Sci. 6: 1–9.
Jonak, C., L. Ökrész, L. Bögre et al. 2002. Complexity, cross talk and integration of plant MAP kinase signaling. Curr Opin Plant Biol. 5(5): 415–424.
Kangasjärvi, S., J. Neukermans, S. Li et al. 2012. Photosynthesis, photorespiration, and light signaling in defence responses. J Exp Bot. 63(4): 1619–1636.
Kästner, J., D. von Knorre, H. Himanshu et al. 2014. Salicylic acid, a plant defense hormone, is specifically secreted by a molluscan herbivore. PloS One. 9(1): e86500.
Kovtun, Y., W.L. Chiu, G. Tena et al. 2000. Functional analysis of oxidative stress-activated mitogen-activated protein kinase cascade in plants. Proc Natl Acad Sci USA. 97: 2940–2945.
Krall, J.P. and G.E. Edwards. 1992. Relationship between photosystem II activity and CO_2 fixation in leaves. Physiol Plantarum. 86(1): 180–187.
Kumar, K. and A.K. Sinha. 2013. Overexpression of constitutively active mitogen activated protein kinase kinase 6 enhances tolerance to salt stress in rice. Rice J. 6: 25.
Lamb, C. and R.A. Dixon. 1997. The oxidative burst in plant disease resistance. Annu Rev Plant Biol. 48(1): 251–275.
Leakey, A.D.B., M.C. Press and J.D. Scholes. 2003. Patterns of dynamic irradiance affect the photosynthetic capacity and growth of dipterocarp tree seedlings. Oecologia. 135(2): 184–193.
Leon, J., M.A. Lawton and I. Raskin. 1995. Hydrogen peroxide stimulates salicylic acid biosynthesis in tobacco. Plant Physiol. 108(4): 1673–1678.
Levine, A., R. Tenhaken, R. Dixon et al. 1994. H_2O_2 from the oxidative burst orchestrates the plant hypersensitive disease resistance response. Cell. 79(4): 583–593.
Levine, E., J.L. Spencer, S.A. Isard et al. 2002. Adaptation of the western corn rootworm to crop rotation: evolution of a new strain in response to a management practice. Am Entomol. 48(2): 94–117.
Li, X.P., A. Phippard, J. Pasari et al. 2002. Structure–function analysis of photosystem II subunit S (PsbS) *in vivo*. Funct Plant Biol. 29(10): 1131–1139.
Lincoln, D.E., D. Couvet and N. Sionit. 1986. Response of an insect herbivore to host plants grown in carbon dioxide enriched atmospheres. Oecologia. 69(4): 556–560.
Lindroth, R.L. 2012. Atmospheric change, plant secondary metabolites and ecological interactions. pp. 120–153. *In*: Iason, G.R., M. Dicke and S.E. Hartley (eds.). The Ecology of Plant Secondary Metabolites: From Genes to Global Processes. Cambridge University Press, Cambridge, UK.
Marino, D., C. Dunand, A. Puppo et al. 2012. A burst of plant NADPH oxidases. Trends Plant Sci. 17(1): 9–15.
Mateo, A., D. Funck, P. Mühlenbock et al. 2006. Controlled levels of salicylic acid are required for optimal photosynthesis and redox homeostasis. J Exp Bot. 57(8): 1795–1807.
Matros, A., S. Amme, B. Kettig et al. 2006. Growth at elevated CO_2 concentrations leads to modified profiles of secondary metabolites in tobacco cv. SamsunNN and to increased resistance against infection with potato virus Y. Plant Cell Environ. 29(1): 126–137.
Mehler, A.H. 1951. Studies on reactions of illuminated chloroplasts: I. Mechanism of the reduction of oxygen and other hill reagents. Arch Biochem Biophys. 33(1): 65–77.
Meldau, S., M. Erb and I.T. Baldwin. 2012. Defence on demand: mechanisms behind optimal defence patterns. Ann Bot-London. 110(8): 1503–1514.
Meza-Canales, I.D., S. Meldau, J.A. Zavala et al. 2017. Herbivore perception decreases photosynthetic carbon assimilation and reduces stomatal conductance by engaging

12-oxo-phytodienoic acid, mitogen-activated protein kinase 4 and cytokinin perception. Plant Cell Environ. 40(7): 1039–1056.
Mhamdi, A. and G. Noctor. 2016. High CO_2 primes plant biotic stress defences through redox-linked pathways. Plant Physiol. DOI:10.1104/pp.16.01129.
Mittler, R., S. Vanderauwera, M. Gollery et al. 2004. Reactive oxygen gene network of plants. Trends Plant Sci. 9: 490–498.
Miyake, C., M. Miyata, Y. Shinzaki et al. 2005. CO_2 response of cyclic electron flow around PSI (CEF-PSI) in tobacco leaves—relative electron fluxes through PSI and PSII determine the magnitude of non-photochemical quenching (NPQ) of Chl fluorescence. Plant Cell Phys. 46(4): 629–637.
Møller, I.M. 2001. Plant mitochondria and oxidative stress: electron transport, NADPH turnover, and metabolism of reactive oxygen species. Annu Rev Plant Biol. 52(1): 561–591.
Munekage, Y., M. Hashimoto, C. Miyake et al. 2004. Cyclic electron flow around photosystem I is essential for photosynthesis. Nature. 429(6991): 579–582.
Musser, R.O., S.M. Hum-Musser, H. Eichenseer et al. 2002. Herbivory: caterpillar saliva beats plant defences. Nature. 416(6881): 599–600.
Nabity, P.D., J.A. Zavala and E.H. DeLucia. 2013. Herbivore induction of jasmonic acid and chemical defences reduce photosynthesis in Nicotiana attenuata. J Exp Bot. 64(2): 685–694.
Niyogi, K.K., A.R. Grossman and O. Björkman. 1998. Arabidopsis mutants define a central role for the xanthophyll cycle in the regulation of photosynthetic energy conversion. Plant Cell. 10(7): 1121–1134.
Niyogi, K.K. 1999. Photoprotection revisited: genetic and molecular approaches. Annu Rev Plant Biol. 50(1): 333–359.
Noctor, G., A. Mhamdi and C.H. Foyer. 2016. Oxidative stress and antioxidative systems: recipes for successful data collection and interpretation. Plant Cell Environ. 39: 1140–1160.
Nurnberger, T., F. Brunner, B. Kemmerling et al. 2004. Innate immunity in plants and animals: striking similarities and obvious differences. Immunol Rev. 198: 249–266.
Ogren, W.L. and G. Bowes. 1971. Ribulose diphosphate carboxylase regulates soybean photorespiration. Nature. 230(13): 159–160.
Ogren, W.L. 1984. Photorespiration: pathways, regulation, and modification. Annu Rev Plant Physiol. 35(1): 415–442.
O'Neill, B.F., A.R. Zangerl, E.H. DeLucia et al. 2008. Longevity and fecundity of Japanese Beetle (*Popillia japonica*) on foliage grown under elevated carbon dioxide. Environ Entomol. 37: 601–607.
O'Neill, B.F., A.R. Zangerl, O. Dermody et al. 2010. Impact of elevated levels of atmospheric CO_2 and herbivory on flavonoids of soybean (Glycine max Linnaeus). J Chem Ecol. 36(1): 35–45.
Ort, D.R. 2001. When there is too much light. Plant Physiol. 125(1): 29–32.
Qiu, Q.S., J.L. Huber, F.L. Booker et al. 2008. Increased protein carbonylation in leaves of Arabidopsis and soybean in response to elevated [CO_2]. Photosynth Res. 97(2): 155–166.
Raven, J.A. 2013. Rubisco: still the most abundant protein of Earth? New Phytol. 198(1): 1–3.
Rivas-San Vicente, M. and J. Plasencia. 2011. Salicylic acid beyond defence: its role in plant growth and development. J Exp Bot. 62(10): 3321–3338.
Rodriguez, M.C., M. Petersen and J. Mundy. 2010. Mitogen-activated protein kinase signaling in plants. Annu Rev Plant Biol. 61: 621–649.
Ruuska, S.A., M.R. Badger, T.J. Andrews et al. 2000. Photosynthetic electron sinks in transgenic tobacco with reduced amounts of Rubisco: little evidence for significant Mehler reaction. J Exp Bot. 51(suppl 1): 357–368.
Schuman, M.C. and I.T. Baldwin. 2016. The layers of plant responses to insect herbivores. Annu Rev Entomol. 61: 373–394.
Seyfferth, C. and K. Tsuda. 2014. Salicylic acid signal transduction: the initiation of biosynthesis, perception and transcriptional reprogramming. Front Plant Sci. 5: 697.

Sheikh, A.H., R. Badmi, S.K. Jalmi et al. 2013. Interaction between two rice mitogen activated protein kinases and its possible role in plant defense. BMC Plant Biol. 13: 121.
Singh, R. and N.S. Jwa. 2013. The rice MAPKK-MAPK interactome: the biological significance of MAPK components in hormone signal transduction. Plant Cell Rep. 32: 923–931.
Tada, Y., S.H. Spoel, K. Pajerowska-Mukhtar et al. 2008. Plant immunity requires conformational charges of NPR1 via S-nitrosylation and thioredoxins. Science. 321(5891): 952–956.
Thaler, J.S., P.T. Humphrey and N.K. Whiteman. 2012. Evolution of jasmonate and salicylate signal crosstalk. Trends Plant Sci. 17(5): 260–270.
Tikkanen, M., P.J. Gollan, N.R. Mekala et al. 2014. Light-harvesting mutants show differential gene expression upon shift to high light as a consequence of photosynthetic redox and reactive oxygen species metabolism. Philos T R Soc B. 369(1640): 20130229.
Torres, M.A., J.D. Jones and J.L. Dangl. 2006. Reactive oxygen species signaling in response to pathogens. Plant Physiol. 141(2): 373–378.
Trewavas, A. 2007. Response to Alpi et al.: Plant neurobiology—all metaphors have value. Trends Plant Sci. 12(6): 231–233.
Vlot, A.C., D.M.A. Dempsey and D.F. Klessig. 2009. Salicylic acid, a multifaceted hormone to combat disease. Annu Rev Phytopathol. 47: 177–206.
Walker, B.J., D.D. Strand, D.M. Kramer et al. 2014. The response of cyclic electron flow around photosystem I to changes in photorespiration and nitrate assimilation. Plant Physiol. 165(1): 453–462.
Wasternack, C. and B. Hause. 2013. Jasmonates: biosynthesis, perception, signal transduction and action in plant stress response, growth and development. An update to the 2007 review in Annals of Botany. Ann Bot-London. 111(6): 1021–1058.
Watling, J.R., M.C. Ball and I.E. Woodrow. 1997. The utilization of light flecks for growth in four Australian rain-forest species. Funct Ecol. 11(2): 231–239.
Weissbach, A., B.L. Horecker and J. Hurwitz. 1956. The enzymatic formation of phosphoglyceric acid from ribulose diphosphate and carbon dioxide. J Biol Chem. 218(2): 795–810.
Woodson, J.D. and J. Chory. 2008. Coordination of gene expression between organellar and nuclear genomes. Nat Rev Gen. 9(5): 383–395.
Zavala, J.A., C.L. Casteel, E.H. DeLucia et al. 2008. Anthropogenic increase in carbon dioxide compromises plant defense against invasive insects. Proc Natl Acad Sci USA. 105(13): 5129–5133.
Zavala, J.A., C.L. Casteel, P.D. Nabity et al. 2009. Role of cysteine proteinase inhibitors in preference of Japanese beetles (*Popillia japonica*) for soybean (*Glycine max*) leaves of different ages and grown under elevated CO_2. Oecol. 161(1): 35–41.
Zavala, J.A., P.D. Nabity and E.H. DeLucia. 2013. An emerging understanding of mechanisms governing insect herbivory under elevated CO_2. Annu Rev Entomol. 58: 79–97.
Zangerl, A.R. and F.A. Bazzaz. 1992. Theory and pattern in plant defense allocation. pp. 363–391. *In*: Fritz, E.L. and E.L. Simms (eds.). Plant Resistance to Herbivores and Pathogens. University of Chicago Press, Chicago, USA.
Zhang, S., X. Li, Z. Sun et al. 2015. Antagonism between phytohormone signalling underlies the variation in disease susceptibility of tomato plants under elevated CO_2. J Exp Bot. 66(7): 1951–1963.
Zheng, X.Y., N.W. Spivey, W. Zeng et al. 2012. Coronatine promotes *Pseudomonas syringae* virulence in plants by activating a signaling cascade that inhibits salicylic acid accumulation. Cell Host Microbe. 11(6): 587–596.

CHAPTER 6

Transgenic Approaches to Combatting Insect Pests in the Field

Model Crops and Recent Environmental and Ecological Friendly Paradigms

Jennifer Campbell, Jason Veizaj, Nicholas Evans, Samantha Taylor and *Chandrakanth Emani**

INTRODUCTION

Globally, the annual agricultural expenditure on insect control amounts to billions of dollars, but this expenditure notwithstanding, 40% of crop losses worldwide are still due to insect pests (Oerke 2006). The second largest order of the insect species, namely Lepidoptera (comprising of the moths and butterflies) alone affects some of the global economic crops such as cotton, tobacco, tomato, corn, sorghum, lucerne, sunflower, pulses, and wheat (Srinivasan et al. 2006). Broad spectrum chemical insecticides were utilized as the primary control agent against insect pests and interestingly, about 40% targeted the lepidopterans (Brooke and Hines 1999). The widespread and sometimes the indiscriminate use of pesticides resulted in the evolution of pesticide resistant insects coupled with reduction in beneficial insect populations (Fitt 1994; Gunning et al. 1991). Synthetic pesticides also had harmful effects on human health and affected the environment in various ways (Gatehouse et al. 1994). These problems have led researchers to focus on environmental friendly strategies that started the revolution of plant genetic engineering with the first commercial transgenic plants expressing plant defense molecules

Department of Biology, Western Kentucky University-Owensboro, 4821 New Hartford Road, Owensboro, KY 42303, USA.
* Corresponding author: chandrakanth.emani@wku.edu

against insect pests. In fact, the first transgenic plant that expressed an insecticidal gene was tobacco plant that produced cowpea trypsin inhibitor for protection against the lepidopteran pest *Heliothis virescens* (Hilder et al. 1987). However, from the model plant tobacco, when this strategy with the same gene was subsequently transferred into rice (Xu et al. 1996) and potato (Gatehouse et al. 1997), it did not provide sustainable insect protection and was thus a commercial dampener. The first commercially successful insect transgenic plants with *Bacillus thuringenesis* (Bt) toxins (Bravo et al. 2007) provided effective protection against lepidopterans in crops such as tobacco (Barton et al. 1987) and tomato (Fischoff et al. 1987). The most commercially viable protection was in cotton (James 2011) that led to increased cotton yields up to 60% in India that translated to decreased insecticide sprays and an annual agricultural income increase of $11.9 billion. Globally, the commercial planting of transgenic crops that was adopted by farmers in 1996, witnessed a revolutionary and dramatic increase in just under two decades and in 2011, 16.7 million farmers in 29 countries planted 160 million hectares of the biotech crops. The global implications of this was seen in the form of a reduction in chemical pesticide use of 443 million Kg which translated into a financial gain of $78 billion (James 2011) for farmers. The present chapter while attempting to showcase the transgenic agricultural revolution looks at specific case studies in economically important crops. The chapter also seeks to review the present novel research areas that focus into relevant future agricultural paradigms such as integrated pest management (that takes into account the ecological implications of insect control) and the current emerging focus on climate change that points to development of climate resilient insect resistant crops.

Transgenic Insect Resistant Crops

Molecular Events of Plant-Insect Pest Interactions—Basis for Transgenic Engineering

A general molecular model that summarizes all the key host-pathogen interaction events in a plant cell attacked by an insect pest essentially accounts for an eight-step signaling pattern in a plant cell (Wu and Baldwin 2010): Step 1 involves the perception of insect pest-derived elicitors by certain unidentified plasma membrane receptors. In step 2, these perception events trigger the activation of Ca^{2+} channels in the form of Ca^{2+} influxes in the plant biochemical realm. In step 3, there occurs a binding of Ca^{2+} to NADPH oxidase resulting in the enzyme's phosphorylation by calcium dependent protein kinases (CDPKs). This transits into the crucial step 4, where the production of reactive oxygen species (ROS) modifies ROS-dependent transcription factors that results in an array of plant defense responses (Lamb and Dixon 1997). Step 5 is a witness to mitogen-activated protein kinases' (MAPKs) activation, among which are salicylic-acid induced protein kinases (SIPKs) and wound-induced protein kinases (WIPKs) that trigger the synthesis of jasmonates. In step 6, the synthesized jasmonates induce an array of molecular responses that inhibit the myelocytomatosis (MYC) transcription factors that in step 7 increase ethylene synthesis. The final step 8 witnesses a series of signaling events induced by elevated ethylene production leading to increased activities of ethylene-responsive transcription factors that are translated into accumulation of metabolites that function as defensive biochemical and molecular entities in plants.

The Evolving Research of Transgenic Insect Resistance

The initial promising biotechnological commercial success stories of insect-resistant *Bacillus thuringenesis* (*Bt*) plants in corn also served as platforms for public and political debate on genetically modified plants. Transgenic *Bt* cotton, maize and rice resistant to a range of lepidopteran pests are now major recognizable agricultural products with well documented insect virulence coupled with an extreme degree of antibiosis (Tabashnik et al. 2003). *Bt* crops served as model systems for crop researchers around the world to assess fitness costs in the fields (Gassman et al. 2009).

On a global scale, insect resistant plants comprise 15% of field-planted transgenic plants (that themselves were 8% of total agricultural crops) (James 2008). Looking through the lens of an ecologist, the transgenic insect resistant plants were considered as vital components of integrated pest management due to their potential characteristic of being environmentally benign while serving as durable pest management crop systems (Kos et al. 2009). Table 1 lists important milestones in engineering transgenic plants resistant to insect pests with the majority of the transgenic plants containing identified and well researched proteins with insecticidal activities (Gatehouse and Gatehouse 1998).

Transgenic insect pest resistant crops were the first practical examples of the successful transfer of biotechnology from lab to the field and in parallel, resulted in a controversial and much debated public and political acceptance of this technology. Recent efforts by responsible scientists are focusing on educating both the farmers and the general public about the economic and environmental benefits of this technology (Gatehouse 2008).

Table 1. Transgenic plants expressing insecticidal plant genes.

Plant	Genes	Insect target	Reference
Cotton	Bt Cry1Ac +	*Heliocarpa zea*	Stewart et al. 2001
	Cry2Ab	*Spodoptera frugiperda*	Chitkowski et al. 2003
		Heliothis virescens	Gahan et al. 2005
Tobacco	CpTI	*Heliothis virescens*	Hilder et al. 1987
	Pot PI II	Lepidoptera	Johnson et al. 1989
	GNA	*Myzus persicae*	Hilder et al. 1995
	p-lec	*H. virescens*	Boulter et al. 1990
	CpTI + p-lec	*H. virescens*	Boulter et al. 1990
	Na PI	*Helicoverpa punctigera*	Heath et al. 1997
	Bt Cry2Aa2	*Helicoverpa armigera*	Kota et al. 1995
	Bt Cry2Aa2	*Helicoverpa armigera*	De Cosa et al. 2001
Potato	CpTI	*Lacanobia oleracea*	Gatehouse et al. 1997
	GNA	*L. oleracea*	Gatehouse et al. 1997
		M. persicae	Gatehouse et al. 1996
		Aulacorthum solani	Down et al. 1996
	GNA + BCH	*M. persicae*	Gatehouse et al. 1996
	Bt Cry1Ba +	*Phthomera opercula*	Naimov et al. 2003
	Cry 1Ia	*Letinotarsa decemlineata*	
		Sesamia inferens	
Rice	Pot PI II	*Chilo suppressalis*	Duan et al. 1996
		Sesamia inferens	Duan et al. 1996
	CpTI	*Chilo suppressalis*	Xu et al. 1996
		Nilaparvata lugens	Xu et al. 1996
	GNA	*Nepphotettix viriscens*	Rao et al. 1998
		Hemiptera	Foissac et al. 2000
	ASA-L		Saha et al. 2006
		Otiorhynchus sulcatus	
Strawberry	CpTI		Graham et al. 1995
		Zabrotes subfaciatus	
Pea	α-AI	*Bruchus pisorum*	Shade et al. 1994
	α-AI		Morton et al. 2000
		Callosobruchus chinensis	
Azuki bean	α-AI		Ishimoto et al. 1996
		Chrysomela tremulae	
Poplar	OC-I	Lepidoptera	Leple et al. 1995
Soybean	Bt Cry1Ab	*Diabrotica virgifera*	Dufourmantel et al. 2005
		Lepidoptera	
Maize	Bt Cry34/35	Rootworm	Moellenbeck et al. 2001
	Bt Cry genes	Coleoptera	Grainnet 2007
	Bt Cry3Bb1	Coleoptera	Vaughn et al. 2005
	Avidin		Kramer et al. 2000
	RNAi ATPase	*Plutella xylostella*	Baum et al. 2007
Broccoli	Bt Cry1Ac/1C		Zhao et al. 2005

Model Examples of Transgenic Insect Resistant Plants

The Diverse Transgenic array of Bt Cotton

Cotton, one of the most economically important crops, owing to its inherent nature of having indeterminate growth traits, renders itself as a vulnerable food and shelter system for about 130 diverse species of insect pests. In designing insect resistant varieties to overcome the diverse limitations, donor genes for insect resistance have been identified from a mind-boggling variety of sources such as bacteria (Perlak et al. 1990), insects themselves (Thomas et al. 1995) and plants (Wu et al. 2006). The rigorous research involving such an array of gene sources led to the pioneer-shifting discovery that had a major role in generating insect resistant transgenic cotton and that led to a revolution in commercial biotechnology, namely, the donor genes encoding crystalline (Cry) proteins of *Bacillus thuringenesis,* more specifically, the Cry δ-endotoxin genes that provided resistance against lepidopteran pests such as bollworm *Helicoverpa zea,* tobacco budworm *Heliothis virescens* (Siebert et al. 2008), pink bollworm *Pectinophora gossypiella* (Tabashnik et al. 2002), and fall armyworm *Spodoptera furgiperda* (Greenplate et al. 2003). A broad-spectrum insect resistant transgenic cotton was developed by engineering the synthetic Cry 1EC protein fused to a tobacco pathogenesis related promoter which showed a 100% mortality against *S. litura* larvae with enhanced expression upon insect bite and salicylic acid treatment (Kumar et al. 2009).

The euphoria of initial successes was short-lived as ecological and evolutionary consequences soon took over the Bt transgenic strategy and insects such as *H. viscerens, P. gossypiella, H. armigera,* and *H. zea* were found to develop resistance against Cry proteins (Bravo and Soberon 2008). The challenges were however overcome when transgenic research employed genes encoding vegetative insecticidal proteins (Vips) that express during the pests' vegetative growth phase starting at mid-log phase and sporulation unlike the regular δ-endotoxins that express only during sporulation (Estruch et al. 1996). The resultant transgenic plants exhibited the ability of an increased resistance range and diversity of action against both lepidopteran pests (Zhu et al. 2006) and coleopteran pests (Wu et al. 2011).

Insect-derived toxins such as *Androctonus austrais* hector insect toxin (AaHIT) resulted in transgenic cotton resistant to cotton bollworm (Wu et al. 2008) and *Manduca sexta* derived protease inhibitors against whitefly *Bemesia tabaci* (Thomas et al. 1995). Transgenic cotton that significantly retarded the larva growth of *H. armigera* was developed by RNAi-mediated suppression of the cotton bollworm P450 monoxygenase gene (*CYP6AE14*) (Mao et al. 2011).

Insecticidal proteins were also identified in plants such as proteases and lectins, specifically, the cowpea (*Vigna unguiculata*) trypsin inhibitor that was genetically engineered in cotton to confer protection against cotton bollworm (Li et al. 1998), lectin derivatives *A. caudatus* agglutinin (*aca*) gene to confer resistance against aphids (Wu et al. 2006) and *A. sativum* agglutinin (*asa*) gene to confer resistance against sap sucking insects (Balogun et al. 2011).

The Multi-pronged Approaches to Develop Insect Resistant Rice

In the model monocot, rice, the array of insect pests that devastated the Asian rice fields included the yellow stem borer (*Scirpophaga incertula*), gall-midge (*Orseolia oryzae*), brown planthopper (*Nilaparvata lugens*), white-backed planthopper (*Sogatella furcifera*), green leafhopper (*Nephotettix virescens*), striped stem borer (*Chilo suppressalis*), rice leaf folder (*Cnaphalocrocis medinalis*). In rice, insects not only acted as primary pests, but also as vectors for viral infections.

The initial attempts in transgenic strategies to develop insect-resistant rice that employed the baculovirus through the insect's enhanced susceptibility by expressing the virus-enhancing factor (Hukuhara et al. 1999) were not effective due to the slow mode of action of the virus. Subsequent strategies that proved more effective employed protease inhibitors targeted against the insect proteases. The initial attempts utilizing protease inhibitors such as the *SBTI, CPTI, PINII,* and *ITR1* met with limited success (Duan et al. 1996; Xu et al. 1996; Lee et al. 1999; Alfonso-Rubi et al. 2003). As in cotton, the most popular and effective approach in developing transgenic insect resistant rice proved to be the use of *Bt Cry* genes. The first transgenic rice expressing the *cry1Ab* (Fujimoto et al. 1993) was swiftly extended to a diverse array of rice varieties engineered with different *cry* proteins in their original, modified, or fully synthetic versions (extensively reviewed in Tyagi et al. 1999; Bajaj and Mohanty 2005; Kathuria et al. 2007). A unique aspect in generating transgenic Bt-rice varieties was gene pyramiding of different cry proteins (Maqbool et al. 2001; Loc et al. 2002; Bashir et al. 2005). Fusion transgenic constructs with cry proteins and carbohydrate-binding moieties of lectins was shown to increase the range of target insects (Mehlo et al. 2005). Field tests of transgenic rice expressing only the cry proteins have shown limited success (Tu et al. 2000; Ye et al. 2001, 2003; Wu et al. 2002; Bashir et al. 2005; Breitler et al. 2004). The fusion protein approach, and more effectively the gene pyramiding not just with cry proteins but also with other insecticidal proteins combined with agronomic deployment strategies as part of the integrated pest management was finally the most effective strategy to develop insect resistant rice plants (Christou et al. 2006; Ferry et al. 2006).

Subsequently, when the cry proteins did not prove effective against the hemipteran insects, plant lectins such as the GNA have been shown to be effective against insects such as brown planthopper (BPH), green leafhopper (GLH), and white-backed planthopper (Rao et al. 1998; Sudhakar et al. 1998; Sun et al. 2002; Nagadhara et al. 2003). The *Allium sativum* leaf lectin was shown to be effective against BPH and GLH (Saha et al. 2006). This approach was given a novel twist with the development of transgenic constructs with *GNA* in combination with the cry proteins (Maqbool et al. 2001; Ramesh et al. 2004) or with *SBTI* (Li et al. 2005) extended the range of target insects to which the transgenic plants were tolerant. Certain other molecules such as the avidin (Yoza et al. 2005) and the spider insecticidal protein (Qiu et al. 2001) have been tested in transgenic rice. Transgenic rice engineered the potato proteinase inhibitor II (*pinII*) gene in Nipponbare, Tainung, and Pi4 varieties exhibited resistance to a major insect pest, pink stem borer (Duan et al. 1996).

The first field trials of *Bt*-transgenic rice were with transgenic Minghui 63 and Shanyou 63, which showed high insect resistance, more importantly, with no reduction in yield towards leaf folder and yellow stem borer (Tu et al. 2001). In the same year, successful field trials were conducted of the transgenic KMD1 line of a Chinese *Japonica* cultivar, Xuishi 11 that showed resistance to eight lepidopteran pests (Shu et al. 2000). KMD1 and related transgenic lines were also unique as they were utilized as insect resistant rice germplasm for hybrid rice production (Wang et al. 2002).

Transgenic Insect Resistant Crops—The Climate Change Paradigm

The Intergovernmental Panel on Climate Change (IPCC) produced important, crucial and well researched documentation relevant to climate change as vital information spanning diverse fields for policy makers and the general public (IPCC 2001). These efforts reached their peak fruition when the organization received the Nobel peace prize (http://www.nobelprize.org/nobel_prizes/peace/laureates/2007/). The updated reports of IPCC in later years focused on the ever-changing environmental changes with emphasis on the increasing carbon dioxide levels, temperatures and ultraviolet levels (IPCC 2007). If witnessed in parallel, the most affected field by such environmental changes is the field of agriculture. It is a known fact that in present times, agriculture with its vital role in creating the present array of food systems across the world is increasingly affected by transforming factors such as population increase, changing income patterns, urbanization and globalization. These factors in turn affect food production, food marketing and food consumption (Von Braun 2007). A new paradigm that agricultural researchers started paying attention to as a vital additional transforming factor is climate change coupled with climate variability, and is now viewed as a threat to the very subsistence of food production and food security (Parry et al. 2004). Around the same time as the IPCC report was being prepared, climate change effects on plant disease epidemiology were examined (Coakley et al. 1999). However, the immense variations observed in pathosystems added to the effects in different locations have been a challenge and in some cases, an inability for plant scientists to lay down a set of general conclusions on the effects of climate change on plant disease (Coakley et al. 1999). Despite such challenges, it can be definitely stated with overwhelming evidence, that climate change has a definite effect on the dynamics of plant disease epidemiology with definite observable effects on alterations in the developmental stages of pathogens, modifying host resistance and changes in the physiology of host-pathogen interactions (Garrett et al. 2006). Recent research reflects efforts on integrating the pests and pathogens into the scenario of climate change (Gregory et al. 2009; Trumble and Butler 2009) that seek to create plant-insect pest interaction-climate change models that provide reliable model parameters, which will aid the development of future strategies to improved insect resistant, climate resilient plant varieties.

Plant-Insect Interactions Viewed Through the Lens of Climate Change

The observed interactions between plants and insects (reviewed in detail by Wu and Baldwin 2010; Smith and Clement 2012) developed immediately after the natural colonization of plants 510 million years ago, and over one million or more insect pests were classified as the causative agents and plant consuming herbivores (Howe and Jander 2008). The resulting evolutionary struggle posited plants to continually evolve mechanisms to minimize the insect consumption while insects evolved mechanisms to overcome the plant defense strategies (Wu and Baldwin 2010; Smith and Clemente 2012). Such evolutionary relationships are now being closely examined in the context of climate change, with the first studies based on observations such as elevated CO_2 and temperature levels altering the interactions between plants and their insect pests (DeLucia et al. 2012).

Current research showed that three major factors of climate change, namely temperature, moisture and CO_2, have an effect on all the interacting disease factors (Pedzoldt and Seaman 2007; Gregory et al. 2009; Trumble and Butler 2009). This sets forth a hypothesis that a plant disease-climate change model encompasses all the intervening aspects of climate and interacting aspects of disease in a mechanistic inclusive fashion (Gregory et al. 2009; DeLucia et al. 2012).

A recent demonstration of specific effects of climate change factors on plant-insect interaction network was in altering the levels of defense hormones such as jasmonic acid and salicylic acid (DeLucia et al. 2012). This specific differential effect on defense biomolecules that results in subsequent changes in plant hormones increases susceptibility of the attacking pests and enhances the plant's innate resistance to pathogens (DeLucia et al. 2012). Such specific studies of the effect of individual factors of climate change and their role in specifically affecting the interactions between plants and insects (see Chapter 5 of this volume) will enable researchers to develop novel strategies to design climate resilient pest resistant plants.

Insect Resistant Crops with Climate Resilience

Insect-pests account for approximately 15% of the world's annual agricultural produce losses that affects both production and food security (Christou 2005). The widespread use of insecticides leads to environmental pollution, human-livestock toxicity and the more dangerous possibility of evolving resistance in insect pests while targeting non-target insects thus offsetting the ecological balance. A solution to this environmental disaster situation came in the form of natural biological control by the use of microbes that evolved into the agricultural innovation known as integrated pest management (IPM) (Trumble 1998). This practice, however, had limited success in large scale agricultural practices especially if viewed in the context of sustainable food production (Bale et al. 2008). A well thought out research partnership between conventional breeders and molecular biotechnologists was a viable integrative option that exploited the plant's inherent resistance, and the genetic and molecular components of the plant-pest interaction mechanisms (Wu and Baldwin 2010). Transgenic technology thus evolved into a complementary alternative strategy to supplement the efforts of plant breeders (Gatehouse 2008; Gatehouse and Gatehouse 1998) and is now a vital integrative component of IPM (Kos et al. 2009). This sustained evolution of biotechnological approaches can now offer a major reassessment of strategies in the contextual global scenario of climate change as farmers and agricultural researchers across the world are now experiencing major impacts of insect management control strategies linked with changes in climate. Thus, it is imperative that present research integrates the pests and pathogens into the climate change/food security debate (Gregory et al. 2009).

The well documented studies related to area-wide suppression of insect-pest control using *Bt* plants (Hutchison et al. 2010) were extended to include climate change parameters. Interactions of elevated CO_2 and nitrogen fertilization effects on the production of *Bt* toxins in transgenic cotton at the field level revealed that CO_2 level elevation from ambient 370 ppm to 900 ppm resulted in reduced *Bt* toxin production and the increased leaf consumption was not compensated even if the plants were treated with increased contents of nitrogen fertilizers (Coviella et al. 2000). To improve the target insect spectrum of *Bt* crops the existing effective strategy of transgene pyramiding was further improved by fusing nontoxic ricin B-chain molecule to *Bt* transgene constructs (Mehlo et al. 2005) for a more sustainable and durable insect pest resistance and the efforts were rewarded by a wider acceptance of this technology in the farming community (Christou 2005). In the current times, such integrative efforts benefit the realm of climate change debates that are also facing the same consequences as the plant biotechnology debates in terms of public and political acceptance.

Integrated Pest Management within the Climate Change Paradigm

Integrated pest management can be termed as a seamless integration of biological controls such as the predators, parasites and pathogens; chemical controls such as the pesticides and cultural controls such as resistant crop varieties and planting times that evolved as focused attempts to reduce insect populations in the field to circumvent economic losses (Trumble and Butler 2009). Many recent studies in the management strategy showed that each of the crucial parameters are directly affected by climate changes, specifically, that modest increases in temperature reduce effectiveness of insect pathogens (Stacy and Fellowes 2002) as well as the pest suppression provided by parasites (Hance et al. 2007). Elevated temperature levels also were found to favor insects with multiple generations in each farming season as opposed to those with a single generation (Bale et al. 2002). Valuable lessons were learnt from the research conducted in insects that are vectors of human pathogens and how their life cycles and spread is interlinked with climate change factors such as temperature and flooding (Trumble and Butler 2009). This enabled crop scientists to effectively redesign the integrated pest management strategies currently being employed by farmers in diverse settings.

Conclusion—Insect Resistance and Food Security within the Climate Change Paradigm

Previous research considered the impact of insect pests on crop yields as a limited global impact factor, but more recent studies have attested the role of pests and diseases as both yield limitation factors as well as early indicators of environmental changes. This is due to their testable variables such as short generation times, high reproductive rates and geographical migration patterns (Scherm et al. 2000). Insects have now become valuable experimental systems in developing models of current and future climate changes by documenting their distribution using climatic mapping in ecological niches (Baker et al. 2000). Insect pests such as aphids that exhibit varied appearance and outbreaks with changing seasons (Gregory et al. 2009) can be used as testable factors for monitoring consequences for agrosystems and agricultural yields. Experimental models that consider agronomic research studies in a climate change context linked to food security policy (Gregory et al. 2009) emphasize that the mechanistic inclusion of pests and disease effects on crop systems will enable realistic predictions of geographically specific crop yields, thus assisting the development of robust regional food security policies. A systematic monitoring of plant-insect pest interactions in a climate change paradigm has revolutionized both conventional breeding and transgenic research in engineering novel genes to generate insect resistant plants. This has enabled revisiting the fundamental concepts of the genetic and molecular bases of plant-insect pest interactions in the light of the effect of climate change as a new focus for research that will determine future policies of food security.

References

Alfonso-Rubi, J., F. Ortego, P. Castanera et al. 2003. Transgenic expression of *trypsin inhibitor CMe* from barley in *Indica* and *Japonica* rice, confers resistance to the rice weevil *Sitophilus oryzae*. Transgenic Res. 12: 23–31.

Bajaj, S. and A. Mohanty. 2005. Recent advances in rice biotechnology—Towards genetically superior transgenic rice. Plant Biotechnol J. 3: 275–307.

Baker, R.H.A., C.E. Sansford, C.H. Jarvis et al. 2000. The role of climatic mapping in predicting the potential geographical distribution of non-indigenous pests under current and future climates. Agri Ecosyst Environ. 82: 57–71.

Bale, J.S., J.C. van Lanteren and F. Bigler. 2008. Biological control and sustainable food production. Phil Trans Roy Soc Lond Sr B. 363: 761–776.

Balogun, N.B., H.M. Inuwa, I. Sani et al. 2011. Expression of mannose-binding insecticidal lectin gene in transgenic cotton (*Gossypium*) plant. Cotton Genomics Genet. 2: 1–7.

Barton, K.A., H.R. Whiteley and N.S. Yang. 1987. *Bacillus thuringiensis* §-endotoxin expressed in transgenic *Nicotiana tabacum* provides resistance to Lepidopteran insects. Plant Physiol. 85: 1103–1109.
Bashir, K., T. Husnain, T. Fatima et al. 2005. Novel indica basmati line (B-370) expressing two unrelated genes of *Bacillus thuringiensis* is highly resistant to two lepidopteran insects in the field. Crop Prot. 24: 870–879.
Baum, J.A., T. Bogaert, W. Clinton et al. 2007. Control of coleopteran insect pests through RNA interference. Nat Biotechnol. 25: 1322–1326.
Boulter, D., G.A. Edwards, A.M.R. Gatehouse et al. 1990. Additive protective effects of different plant derived insect resistance genes in transgenic tobacco plants. Crop Prot. 9: 351–354.
Bravo, A., S.S. Gill and M. Soberón. 2007. Mode of action of *Bacillus thuringiensis* Cry and Cyt Toxins and their potential for insect control. Toxicon. 49: 423–435.
Bravo, A. and M. Soberon. 2008. How to cope with insect resistance to Bt toxins? Trends Biotechnol. 26: 573–579.
Breitler, J.C., D. Meynard, J. Van Boxtel et al. 2004. A novel two T-DNA binary vector allows efficient generation of marker free transgenic plants in three elite cultivars of rice (*Oryza sativa* L.). Transgenic Res. 13: 271–287.
Brooks, E.M. and E.R. Hines. 1999. Viral biopesticides for heliothine control. Fact or faction. Today's Life Sci Jan/Feb: 38–44.
Chitkowski, R.L., S.G. Turnipseed, M.J. Sullivan et al. 2003. Field and laboratory evaluations of transgenic cottons expressing one or two *Bacillus thuringiensis* var. kurstaki Berliner proteins for management of noctuid (Lepidoptera) pests. J Econ Entomol. 96: 755–762.
Christou, P. 2005. Sustainable and durable insect pest resistance in transgenic crops: http://www.isb.vt.edu/news/2005/artspdf/aug0503.pdf.
Christou, P., T. Capell, A. Kohli et al. 2006. Recent developments and future prospects in insect pest control in transgenic crops. Trends Plant Sci. 11: 302–308.
Coakley, S.M., H. Scherm and S. Chakraborthy. 1999. Climate change and plant disease management. Ann Rev Phytopath. 37: 399–426.
Coviella, C.E., D.J. Morgan and J.T. Trumble. 2000. Interactions of elevated CO_2 and nitrogen fertilization: effects on production of *Bacillus thuringenesis* toxins in transgenic plants. Environ Entomol. 29: 781–787.
De Cosa, B., W. Moar, S.B. Lee et al. 2001. Overexpression of the Bt cry2Aa2 operon in chloroplasts leads to formation of insecticidal crystals. Nat Biotechnol. 19: 71–74.
De Lucia, E., P. Nabity, J. Zavala et al. 2012. Climate change: Resetting plant-insect interactions. Plant Physiol. 160: 1677–1685.
Down, R.E., A.M.R. Gatehouse, W.D.O. Hamilton et al. 1996. Snowdrop lectin inhibits development and decreases fecundity of the glasshouse potato aphid (*Aulacorthum solani*) when administered *in vivo* and via transgenic plants both in laboratory and glasshouse trials. J Insect Physiol. 42: 1035–1045.
Duan, X., X. Li, Q. Xue et al. 1996. Transgenic rice plants harboring an introduced potato proteinase inhibitor II gene are insect resistant. Nat Biotechnol. 14: 494–498.
Dufourmantel, N., G. Tissot, F. Goutorbe et al. 2005. Generation and analysis of soybean plastid transformants expressing *Bacillus thuringiensis* Cry1Ab protoxin. Plant Mol Biol. 58: 659–668.
Estruch, J.J., G.W. Warren, M.A. Mullins et al. 1996. Vip3A, a novel *Bacillus thuringiensis* vegetative insecticidal protein with a wide spectrum of activities against lepidopteran insects. Proc Natl Acad Sci USA. 93: 5389–5394.
Ferry, N., M. Edwards, J. Gatehouse et al. 2006. Transgenic plants for insect pest control: a forward looking scientific perspective. Transgenic Res. 15: 13–19.
Fischhoff, D.A., K.S. Bowdish, F.J. Perlak et al. 1987. Insect tolerant transgenic tomato plants. Bio/Technol. 5: 807–813.
Fitt, G.P. 1994. Cotton pest management: part 3. An australian perspective. Annu Rev Entomol. 39: 532–562.

Foissac, X., N.T. Loc, P. Christou et al. 2000. Resistance to green leafhopper (*Nephotettix virescens*) and brown planthopper (*Nilaparvata lugens*) in transgenic rice expressing snowdrop lectin (*Galanthus nivalis* agglutinin; GNA). J Insect Physiol. 46: 573–583.

Fujimoto, H., K. Itoh, M. Yamamoto et al. 1993. Insect resistant rice generated by introduction of a modified delta-endotoxin gene of *Bacillus thuringiensis*. Bio/Technology. 11: 1151–1155.

Gahan, L.J., Y.T. Ma, M.L.M. Coble et al. 2005. Genetic basis of resistance to Cry1Ac and Cry2Aa in *Heliothis virescens* (Lepidoptera: Noctuidae). J Econ Entomol. 98: 1357–1368.

Garrett, K.A., S.P. Dendy, E.E. Frank et al. 2006. Climate change effects on plant disease: Genomes to ecosystems. Annu Rev Phytopathol. 44: 489–509.

Gassmann, A.J., Y. Carriere and B.E. Tabashnik. 2009. Fitness costs of insect resistance to *Bacillus thuringiensis*. Annu Rev Entomol. 54: 147–63.

Gatehouse, J.A. 2008. Biotechnological prospects for engineering insect-resistant plants. Plant Physiol. 146: 881–887.

Gatehouse, A.M.R., R.E. Down, K.S. Powell et al. 1996. Transgenic potato plants with enhanced resistance to the peach-potato aphid *Myzus persicae*. Entomol Exp Appl. 34: 295–307.

Gatehouse, A.M.R., G.M. Davison, C.A. Newell et al. 1997. Transgenic potato plants with enhanced resistance to the tomato moth, *Lacanobia oleracea*: growth room trials. Mol Breed. 3: 49–63.

Gatehouse, A.M.R. and J.A. Gatehouse. 1998. Identifying proteins with insecticidal activity: use of encoding genes to produce insect-resistant transgenic crops. Pestic Sci. 52: 165–175.

Gatehouse, A.M., V.A. Hilder, K.S. Powell et al. 1994. Insect-resistant transgenic plants: choosing the gene to do the 'job'. Biochem Soc Trans. 22: 944–949.

Graham, J., R.J. McNicol and K. Greig. 1995. Towards genetic based insect resistance in strawberry using the cowpea trypsin inhibitor. Ann Appl Biol. 127: 163–173.

Grainnet. 2007. Monsanto and Dow Agrosciences launch "SmartStax", industry's first-ever eight-gene stacked combination in corn. Grainnet: http://www.grainnet.com/.

Greenplate, J.T., J.W. Mullins, S.R. Penn et al. 2003. Partial characterization of cotton plants expressing two toxin proteins from *Bacillus thuringiensis*: relative toxin contribution, toxin interaction, and resistance management. J Appl Ent. 127: 340–347.

Gregory, P.J., S.N. Johnson, J.C. Newton et al. 2009. Integrating pests and pathogens into the climate change/food security debate. J Exp Bot. 60: 2827–2838.

Gunning, R.V., C.S. Easton, M.E. Balfe et al. 1991. Pyrethroid resistance mechanisms in Australian *Helicoverpa armigera*. Pestic Sci. 33: 473–490.

Hance, T., J. van Baaren, P. Vernon et al. 2007. Impact of extreme temperatures on parasitoids in a climate change perspective. Annu Rev Entomol. 52: 107–26.

Heath, R., G. McDonald, J.T. Christeller et al. 1997. Proteinase inhibitors from *Nicotiana alata* enhance plant resistance to insect pests. J Insect Physiol. 833–842.

Hilder, V.A., A.M.R. Gatehouse, S.E. Sherman et al. 1987. A novel mechanism for insect resistance engineered into tobacco. Nature. 330: 160–163.

Hilder, V.A., K.S. Powell, A.M.R. Gatehouse et al. 1995. Expression of snowdrop lectin in transgenic tobacco plants results in added protection against aphids. Transgen Res. 4: 18–25.

Howe, G.A. and G. Jander. 2008. Plant immunity to insect herbivores. Annu Rev Plant Biol. 59: 41–66.

Hukuhara, T., T. Hayakawa and A. Wijonarko. 1999. Increased baculovirus susceptibility of armyworm larvae feeding on transgenic rice plants expressing an entomopoxvirus gene. Nat Biotechnol. 17: 1122–1124.

Hutchison, W.D., E.C. Burkness, P.D. Mitchell et al. 2010. Areawide suppression of European corn borer with *Bt* maize reaps savings to non-*Bt* maize growers. Science. 330: 222–225.

Intergovernmental Panel on Climate Change. 2007. Climate change 2007: The Physical Science Basis. Summary for Policymakers. Contribution of working group I to the third assessment report of the IPCC. IPCC Secretariat, Geneva, Switzerland.

Ishimoto, M., T. Sato, M.J. Chrispeels et al. 1996. Bruchid resistance of transgenic azuki bean expressing seed a-amylase inhibitor of common bean. Entomol Exp Appl. 79: 309–315.
James, C. 2008. Global Status of Commercialized Biotech/GM Crops: 2008 (ISAAA Brief 39), International Service for the Acquisition of Agri-biotech Applications (summary available at http://isaaa.org/resources/publications/briefs/39/download/isaaa-brief-39-2008.pdf).
James, C. 2011. Global status of Commercialized Biotech/GM Crops. ISAAA Brief No. 43 Ithaca, NY, USA.
Johnson, R., J. Narvaez, G. An et al. 1989. Expression of proteinase inhibitors I and II in transgenic tobacco plants: Effects on natural defense against *Manduca sexta* larvae. Proc Nat Acad Sci USA. 86: 9871–9875.
Kathuria, H., J. Giri, H. Tyagi et al. 2007. Advances in transgenic rice biotechnology. Crit Rev Plant Sci. 26: 65–103.
Kos, M., J.J.A. van Loon, M. Dicke et al. 2009. Transgenic plants as vital components of integrated pest management. Trends Biotechnol. 27: 621–627.
Kota, M., H. Daniell, S. Varma et al. 1999. Overexpression of the *Bacillus thuringiensis* (*Bt*) Cry2Aa2 protein in chloroplasts confers resistance to plants against susceptible and Bt-resistant insects. Proc Natl Acad Sci USA. 96: 1840–1845.
Kramer, K.J., T.D. Morgan, J.E. Throne et al. 2000. Transgenic avidin maize is resistant to storage insect pests. Nat Biotechnol. 18: 670–674.
Kumar, M., A.K. Shukla, H. Singh et al. 2009. Development of insect resistant transgenic cotton lines expressing cry1EC gene from an insect bite and wound inducible promoter. J Biotechnol. 140: 143–148.
Lamb, C. and R.A. Dixon. 1997. The oxidative burst in plant disease resistance. Annu Rev Plant Physiol Plant Mol Biol. 48: 251–75.
Lee, S.I., S.H. Lee, J.C. Koo et al. 1999. *Soybean Kunitz trypsin inhibitor* (SKTI) confers resistance to the brown planthopper (*Nilaparvata lugens Stal*) in transgenic rice. Mol Breed. 5: 1–9.
Leple, J.C., M. Bonade-Bottino, S. Augustin et al. 1995. Toxicity to *Chrysomela tremulae* (Coleoptera: *Chrysomelidae*) of transgenic poplars expressing a cysteine proteinase inhibitor. Mol Breed. 1: 319–328.
Li, G.Y., X.P. Xu, H.T. Xing et al. 2005. Insect resistance to *Nilaparvata lugens* and *C. medinalis* in transgenic *Indica* rice and the inheritance of *gna* plus *sbti* transgenes. Pest Management Sci. 61: 390–396.
Li, Y.E., Z. Zhu, Z.X. Chen et al. 1998. Obtaining transgenic cotton plants with cowpea trypsin inhibitor. Acta Goss Sini. 10: 237–243.
Loc, N.T., P. Tinjuangjun, A.M.R. Gatehouse et al. 2002. Linear transgene constructs lacking vector backbone sequences generate transgenic rice plants which accumulate higher levels of proteins conferring insect resistance. Mol Breed. 9: 231–244.
Mao, Y.B., X.Y. Tao, X.Y. Xue et al. 2011. Cotton plants expressing CYP6AE14 double-stranded RNA show enhanced resistance to bollworms. Transgenic Res. 20: 665–673.
Maqbool, S.B., S. Riazuddin, N.T. Loc et al. 2001. Expression of multiple insecticidal genes confers broad resistance against a range of different rice pests. Mol Breed. 7: 85–93.
Mehlo, L., D. Gahakwa, P.-T. Nghia et al. 2005. An alternative strategy for sustainable pest resistance in genetically enhanced crops. Proc Natl Acad Sci USA. 102: 7812–7816.
Moellenbeck, D.J., M.L. Peters, J.W. Bing et al. 2001. Insecticidal proteins from *Bacillus thuringiensis* protect corn from corn rootworms. Nat Biotechnol. 19: 668–672.
Morton, R.L., H.E. Schroeder, K.S. Bateman et al. 2000. Bean alpha-amylase inhibitor 1 in transgenic peas (*Pisum sativum*) provides complete protection from pea weevil (*Bruchus pisorum*) under field conditions. Proc Natl Acad Sci USA. 97: 3820–825.
Naimov, S., S. Dukiandjiev and R.A. de Maagd. 2003. A hybrid *Bacillus thuringiensis* delta-endotoxin gives resistance against a coleopteran and a lepidopteran pest in transgenic potato. Plant Biotechnol J. 1: 51–57.
Nagadhara, D., S. Ramesh, I.C. Pasalu et al. 2003. Transgenic *Indica* rice resistant to sap-sucking insects. Plant Biotechnol J. 1: 231–240.

Oerke, E.C. 2006. Crop losses to pests. J Agric Sci. 144: 31–43.

Parry, M.L., C. Rosenzweig, A. Iglesias et al. 2004. Effects of climate change on global food production under SRES emissions and socio-economic scenarios. Glob Environ Change. 14: 53–67.

Perlak, F.J., R.W. Deaton, T.A. Armstrong et al. 1990. Insect resistant cotton plants. Bio/Technol. 8: 939–943.

Petzoldt, C. and A. Seaman. 2007. Climate change effects on insects and pathogens. Fact Sheet: http://www.climateandfarming.org/pdfs/FactSheets/III.2Insects.Pathogens.pdf.

Qiu, H., Z. Wei, H. An et al. 2001. *Agrobacterium tumefaciens* mediated transformation of rice with the spider insecticidal gene conferring resistance to leaf folder and striped stem borer. Cell Res. 11: 149–155.

Ramesh, S., D. Nagadhara, V.D. Reddy et al. 2004. Production of transgenic *Indica* rice resistant to yellow stem borer and sap-sucking insects, using super-binary vectors of *Agrobacterium tumefaciens*. Plant Sci. 166: 1077–1085.

Rao, K.V., K.S. Rathore, T.K. Hodges et al. 1998. Expression of snowdrop lectin (GNA) in transgenic rice plants confers resistance to rice brown planthopper. Plant J. 15: 469–477.

Saha, P., P. Majumder, I. Dutta et al. 2006. Transgenic rice expressing *Allium sativum* leaf lectin with enhanced resistance against sap-sucking insect pests. Planta. 223: 1329–1343.

Scherm, H., R.W. Sutherst, R. Harrington et al. 2000. Global networking for assessment of impacts of global change on plant pests. Environ Poll. 108: 333–341.

Shade, R.E., H.E. Schroeder, J.J. Pueyo et al. 1994. Transgenic peas expressing the a-amylase inhibitor of the common bean are resistant to bruchid beetles. Bio/Technol. 12: 793–796.

Shu, Q., G. Ye, H. Cui et al. 2000. Transgenic rice plants with a synthetic *cry1Ab* gene from *Bacillus thuringiensis* were highly resistant to eight lepidopteran rice pest species. Mol Breed. 6: 433–439.

Siebert, M.W., S. Nolting, B.R. Leonard et al. 2008. Efficacy of transgenic cotton expressing Cry1Ac and Cry1F insecticidal protein against heliothines (Lepidoptera: Noctuidae). J Econ Entomol. 101: 1950–1959.

Smith, C.M. and S.L. Clement. 2012. Molecular bases of plant resistance to arthropods. Annu Rev Entomol. 57: 309–32810.

Srinivasan, A., A. Giri and V. Gupta. 2006. Structural and functional diversities in lepidopteran serine proteases. Cell Mol Biol Lett. 11: 132–154.

Stacey, D.A. and M.D.E. Fellowes. 2002. Influence of temperature on pea aphid *Acyrthosiphon pisum* (Hemiptera: Aphididae) resistance to natural enemy attack. Bull Entomol Res. 92: 351–357.

Stewart, S.D., J.J. Adamczyk, K.S. Knighten et al. 2001. Impact of Bt cottons expressing one or two insecticidal proteins of *Bacillus thuringiensis* Berliner on growth and survival of noctuid (Lepidoptera) larvae. J Econ Entomol. 94: 752–760.

Sudhakar, D., X.D. Fu, E. Stoger et al. 1998. Expression and immunolocalisation of the snowdrop lectin, GNA in transgenic rice plants. Transgenic Res. 7: 371–378.

Sun, X.L., A. Wu and K. Tang. 2002. Transgenic rice lines with enhanced resistance to the small brown planthopper. Crop Prot. 21: 511–514.

Tabashnik, B.E., T.J. Dennehy, M.A. Sims et al. 2002. Control of resistant pink bollworm (*Pectinophora gossypiella*) by transgenic cotton that produces *Bacillus thuringiensis* toxin Cry2Ab. Appl Environ Microbiol. 68: 3790–3794.

Tabashnik, B.E., Y. Carriere, T.J. Dennehy et al. 2003. Insect resistance to transgenic *Bt* crops: lessons from the laboratory and field. J Econ Entomol. 96: 1031–38.

Thomas, J.C., D.G. Adams, V.D. Keppenne et al. 1995. Protease inhibitors of *Manduca sexta* expressed in transgenic cotton. Plant Cell Rep. 14: 758–762.

Trumble, J.T. 1998. IPM: Overcoming conflicts in adoption. Integr Pest Manage Rev. 3: 195–207.

Trumble, J.T. and C.D. Butler. 2009. Climate change will exacerbate California's insect pest problems. Cal Agri. 63: 73–78.

Tu, J., G. Zhang, K. Datta et al. 2000. Field performance of transgenic elite commercial hybrid rice expressing *Bacillus thuringiensis* delta-endotoxin. Nat Biotechnol. 18: 1101–1104.

Tyagi, A.K., A. Mohanty, S. Bajaj et al. 1999. Transgenic rice: a valuable monocot system for crop improvement and gene research. Crit Rev Biotechnol. 19: 41–79.

Vaughn, T., T. Cavato, G. Brar et al. 2005. A method of controlling corn rootworm feeding using a *Bacillus thuringiensis* protein expressed in transgenic maize. Crop Sci. 45: 931–938.

Von Braun, J. 2007. The world food situation: new driving forces and required actions. International Food Policy Research Institute, Washington DC, USA.

Wang, Z., Q. Shu, G. Ye et al. 2002. Genetic analysis of resistance of Bt rice to stripe borer (*Chilo supperessalis*). Euphytica. 123: 379–386.

Wu, J., X. Luo, H. Guo et al. 2006. Transgenic cotton, expressing *Amaranthus caudatus* agglutinin, confers enhanced resistance to aphids. Plant Breed. 125: 390–394.

Wu, J., X. Luo, Z. Wang et al. 2008. Transgenic cotton expressing synthesized scorpion insect toxin AaHIT gene confers enhanced resistance to cotton bollworm (*Heliothis armigera*) larvae. Biotechnol Lett. 30: 547–554.

Wu, J. and I.T. Baldwin. 2010. New insights into plant responses to the attack from insect herbivores. Annu Rev Genet. 44: 1–24.

Wu, J., X. Luo, X. Zhang et al. 2011. Development of insect resistant transgenic cotton with chimeric TVip3A accumulating in chloroplasts. Transgenic Res. 20: 963–973.

Wu, D.X., Q.Y. Shu, Z.H. Wang et al. 2002. Quality variations in transgenic rice with a synthetic *cry1Ab* gene from *Bacillus thuringiensis*. Plant Breeding. 121: 198–202.

Xu, D.P., Q.Z. Xue, D. McElroy et al. 1996. Constitutive expression of a cowpea trypsin-inhibitor gene, CpTI, in transgenic rice plants confers resistance of two major rice insect pests. Mol Breed. 2: 167–173.

Ye, G., Q. Shu, H. Yao et al. 2001. Field evaluation of resistance of transgenic rice containing a synthetic *cry1Ab* gene from *Bacillus thuringiensis* berliner to two stem borers. J Econ Entomol. 94: 271–276.

Yoza, K., T. Imamura, K.J. Kramer et al. 2005. Avidin expressed in transgenic rice confers resistance to the stored-product insect pests *Tribolium confusum* and *Sitotroga cerealella*. Biosci Biotechnol Biochem. 69: 966–971.

Zhao, J.Z., J. Cao, H.L. Collins et al. 2005. Concurrent use of transgenic plants expressing a single and two *Bacillus thuringiensis* genes speeds insect adaptation to pyramided plants. Proc Natl Acad Sci USA. 102: 8426–8430.

CHAPTER 7

Metabolomics of Plant Resistance to Insects

Mirka Macel[1,4] and *Nicole M. van Dam*[2,3,4,*]

INTRODUCTION

Plants have evolved effective defense mechanisms against a plethora of attackers, ranging from microbial pathogens to large mammalian herbivores. These defenses can be mechanistic (hairs, thorns) or chemical of nature. Plant chemical defenses are mostly plant secondary metabolites such as alkaloids, glucosinolates or phenolic compounds. In total, it has been estimated that plants produce over 200,000 different secondary metabolites, many of which play a role in plant herbivore defense (Weckwerth 2003). Because of the overwhelming diversity of compounds, large-scale chemical-analytical screening approaches, such as metabolomics, have increased in popularity for herbivore resistance research. In this chapter, we focus on the chemistry behind plant resistance to insect herbivores and the use of metabolomics to investigate these plant chemical defenses. We highlight the benefits and disadvantages of metabolomics approaches and summarize recent studies using metabolomics in insect resistance studies. Finally, we discuss how metabolomics can be merged with genetic analyses that are currently in use in breeding programs and how this will promote future implementation of systems biological approaches in research programs for insect resistance in crops.

[1] Leiden University, Institute of Biology Leiden, Plant Ecology and Phytochemistry, P.O. Box 9505, 2300 RA Leiden, The Netherlands.
Email: mirkamacel@gmail.com

[2] German Centre for Integrative Biodiversity Research (iDiv) Halle-Jena-Leipzig, Deutscher Platz 5e, 04103 Leipzig, Germany.

[3] Friedrich Schiller University Jena, Institute of Biodiversity, Dornburger-Str. 159, 07743 Jena, Germany.

[4] Radboud University, Molecular Interaction Ecology, Institute of Water and Wetland Research (IWWR), PO Box 9010, 6500 GL Nijmegen, The Netherlands.

* Corresponding author: nicole.vandam@idiv.de

Plant Chemical Defenses

Secondary metabolites that may serve as plant defenses are found in all plant species and plant organs. The production and distribution of individual compounds in the plant, however, is very diverse and their concentrations can greatly vary between and among plant tissues as well as among plant genotypes (Gong et al. 2013; Soltis and Kliebenstein 2015). In addition, plant defense chemistry is species-specific, whereby each species has its own particular blend of plant secondary metabolites (Schweiger et al. 2014; Macel et al. 2014). Plant chemical defenses can be deterrent or toxic to generalist insect herbivores. On the other hand, specialist insects are often adapted to the specific plant secondary metabolites of their host plant and specific compounds may even make the plant more attractive to specialist herbivores (Schoonhoven et al. 2005).

Next to constitutive defenses, plants can also induce the expression of chemical defenses in response to attack by insect herbivores. This induction response may be immediate, although there may be some delay before the defense level has increased substantially, for example for the production of defences such as glucosinolates and trichomes in herbivore-damaged *Brassica juncea* (Mathur et al. 2014). Often, if not always, the induced response is not restricted to the site of attack, but is systemically expressed throughout the whole plant. The levels of chemical defenses may remain higher for at least a week after insect herbivore attack before they drop to lower, pre-attack levels again (Karban and Baldwin 1997).

Due to these temporal dynamics and the systemic nature of the response, attack by one insect herbivore may influence the plants' resistance to another herbivore, even if it arrives later or feeds on other organs of the same plant. For example, attack by root herbivores can lead to changes in plant chemical defense in the shoot which will then affect insects feeding or ovipositing on the shoot (Erb et al. 2008). The effect of belowground herbivore on shoot chemistry can also affect higher trophic levels such as the predators or parasitoids of the herbivores (Soler et al. 2007). In addition, soil microbes as well as abiotic conditions such as drought or flooding, can affect shoot plant chemistry and herbivore resistance (Hol et al. 2010; Nguyen et al. 2016). The effect of these multiple interactions on plant resistance to insect herbivores should be considered when studying chemical defense mechanisms or breeding for sustainable resistance to a wide arrange of herbivores.

Plant Metabolomics

In general, plant defenses have been studied using targeted approaches, for example by analyzing plant alkaloids and their effects on insects. Plant metabolomics offers a new approach. The metabolome of an organism has been defined as the total of all its metabolites (Fiehn et al. 2000). Metabolomics aims to investigate a wide range of these metabolites that are of various compound classes and hence have a wide range of chemical properties. General analytical techniques that are used for metabolomics are Nuclear Magnetic Resonance (NMR) and Mass-Spectrometry (MS) based techniques, which we will briefly discuss below.

There are two types of metabolomics approaches; untargeted or global metabolomics and targeted metabolomics (Cajka and Fiehn 2016). Untargeted metabolomics is a comprehensive screening of as many metabolites as possible, with unknown as well as known indentities. The emphasis with untargeted metabolomics is on qualification ("fingerprinting") rather than quantification. This can lead to novel insights in mechanisms of plant resistance to insects (e.g., Leiss et al. 2009). "Targeted metabolomics", on the other hand, aims at the quantification of a large set of known metabolites. In contrast to the traditional targeted approaches, which cover only one compound class (e.g., glucosinolates or alkaloids), targeted metabolomics offers the possibility to analyze many known compounds from different compound classes at the same time (Cajka and Fiehn 2016). The advantage of targeted over untargeted metabolomics is that quantification is more exact as proper—preferably labelled—references can be chosen (Cajka and Fiehn 2016). Ultimately, for any given compound or compound class, a targeted (non-metabolomics) procedure which is fine-tuned towards the specific metabolites will give the most reliable quantification. Hence, if the species' defensive metabolites are already known or partly known, the use of either targeted metabolomics or targeted extractions/analyses of these defense compounds may be more useful as they will provide better quantification of these known compounds with known functions. However, if for a given plant species the chemical defenses are still largely unknown, metabolomics is a great start to search for active compounds in plant resistance. For chemical markers of resistance, for example for large-scale phenotypic screening for resistant genotypes, it is not necessary to know the exact function or identity of a metabolite as long as there is a correlation with resistance that can be selected for.

Getting from Metabolites to Functions

With the current metabolomics approaches, detecting a large number of compounds—or (mass features) as peaks representing unidentified compound are called—in a plant sample is relatively easy. However, for many metabolites, even those in model plant species, the physiological or ecological function for the plant is still unknown (Weckwerth 2003).

Therefore, untargeted metabolomics usually not only yields metabolites with unknown identity, but also known (to compound class level) metabolites but with an unknown function (Marti et al. 2013). The elucidation of metabolite identity and especially function can be a long-winded road. Two generally used approaches are bioassay driven fractionation and genetic transformation. Bioassay driven fractionation is a process in which a global plant extract, containing a large part of the metabolome, is fractionated based on polarity (Prince and Pohnert 2010). Each faction is then tested for remaining activity using a bioassay involving the target organism, after which the active fraction is retained and fractionated further until the active compounds can be identified. Whereas the chemical procedures in the process are quite straightforward, bioassay guided fractionation is mostly limited to herbivores that can be reared on artificial diets. In addition, it is quite labor intensive. Moreover, in the cause of the fractionation process, the biological activity may get lost when two compounds showing synergistic effects accidentlly end up in different fractions. Genetic modification, gene silencing via RNAi, or genome editing via CRISPR/Cas9 are other approaches that may be applied, for example to study the effect of—particular classes of—compounds in resistance against insect herbivores (Beekwilder et al. 2008; Weeks et al. 2016). However, genetic modification can only be applied if the genes involved in the biosynthesis of the compounds of interest are known—which is usually only the case for model plants and well-studied compounds classes such as glucosinolates and alkaloids. Even though genetic modification is targeting a single compound (class), the modification itself may also affect other (non-related) plant compounds (Shepherd et al. 2015; Simo et al. 2014). Preferably, such unintended changes are identified using metabolomics before it is concluded that the compounds of which the synthesis is modified specifically causes the observed effects (Simo et al. 2014). Moreover, it should also be considered that in the long run a wide spread use of genetically modified crops may lead to other problems, such as rapid evolution of insect tolerance to the modification (Grassmann et al. 2011).

Metabolomics for Insect Resistance—A Review

As resistance to insect pests is often chemically based, metabolomics has been rapidly adapted as a tool to explore plant species and varieties for hitherto unknown sources for sustainable resistance. Table 1 shows an overview of studies on plant-insect interactions that used a metabolomics approach. The lists include crops as well as wild plants and trees. One of the first studies that used metabolomics to identify insect resistance in crops was performed on chrysantemum (Leiss et al. 2009a). Despite more

Table 1. Untargeted metabolomics studies of plant-insect interactions. Only studies using both plant and insects or insect oral secretions were included. Sorted alphabetically on plant species name.
C = constitutive, I = induced, 3rd = third trophic level, HIPVs = herbivore induced plant volatiles.

Metabolomics platform	**Plant species**	**Type of defense**	**Insect herbivore/ other organisms**	**Relevant metabolites found**	**Assessment**	**Reference**
GC-MS	*Arabidopsis thaliana*	Shoot	Soil microbes, *Trichoplusia ni*	amino acids	Above-belowground interactions	Badri et al. 2013
UHPLC-qTOF-MS	*A. thaliana*	Shoot	*Pseudomonas fluorescens*, *Spodoptera exigua*	salicylic acid, indole glucosinolates	Above-belowground interactions/ Transcriptome-metabolome	Van de Mortel et al. 2012
UHPLC-TOF-MS	*A. thaliana*	I, Root and Shoot	*Heterodera schachtii*, *Brevicoryne brassicae*	Glucosinolates	Above-belowground interactions	Kutyniok and Müller 2012
LC-MS	*Barbarea vulgaris*	C, Shoot	*Phyllotreta nemorum*	Saponins	Genotype comparison/ QTLs-metabolome	Kuzina et al. 2009 Kuzina et al. 2011
GC-MS	*Brassica nigra*	I, Shoot	*Pieris brassiceceae*, *Trichogramma brassicae* (3rd), *Cotesia glomerata* (3rd)	HIPVs	Tritrophic interactions	Fatouros et al. 2012
UPLCT-MS	*Brassica oleracaea*	Root and shoot	*Pieris rapae*	Coumaroylquinic acids	Plant-insect interface	Jansen et al. 2009

Table 1 contd. ...

...Table 1 contd.

Metabolomics platform	Plant species	Type of defense	Insect herbivore/ other organisms	Relevant metabolites found	Assessment	Reference
^{1}H NMR	*Brassica rapa*	I, Shoot	*Plutella xylostella, Spodoptera exigua*	Amino acids, sugars, phenolics, glucosinolates	Control vs. induced	Widarto et al. 2006
^{1}H NMR	Carrot (*Daucus carota*)	C, shoot	*Frankliniella occidentalis*	Amino acids, phenolics	Genotype/ cultivar comparison	Leiss et al. 2013
^{1}H NMR	Chrysanthemum	C, Shoot	*Frankliniella occidentalis*	Chlorogenic acid	Cultivar comparison	Leiss et al. 2009
GC-MS LC-MS	*Citrus aurantium, Citrus reshni*	I, Shoot	*Tetranychus urticae* (mite)	HIPVs, Alkaloids, Flavonoids	Species comparison/ Control vs. induced vs. neighbors/Gene expression	Agut et al. 2015
CE-MS	*Rumex obtusifolius, Foeniculum vulgare*	Shoot	*Gastrophysa atrocyanea, Papilio machaon*	Lactate	Plant-insect interface	Miyagi et al. 2013
GC-MS	Forest communities	Honeydew	Scale insect	Amino acids, carbohydrate	Insect species comparison	Dhami et al. 2011
Targeted LC-MS GC-MS	*Gingko biloba*	I, shoot	*Spodoptera littoralis*	Flavonoids, HIPVs	Control vs. induced/Gene expression	Mohanta et al. 2012
UHPLC-TOF-MS	Maize (*Zea mays*)	I, root and shoot	*Spodoptera littoralis*	Many	Control vs. induced	Marti et al. 2014

LC-MS	Maize	I, shoot	*Rhopalosiphum maidis* (aphid)	Salicylic acid, terpenes	Control vs. induced/ Transcriptome-Metabolome	Tzin et al. 2015
UPLC-MS	Maize	Root and shoot	*Spodoptera frugiperda*, *Spodoptera littoralis*	HMDBOA-Glc	Plant-insect interface	Glauser et al. 2011
Targeted HPLC GC-MS	Mountain Birch *(Betula pubescens)*	I, shoot	*Epirrita autumnata (geometrid)*	Phenolics, amino acids, sugars, organic acids, tocopherol	Different levels of insect attack	Ossipov et al. 2014
UHPLC-MS-qTOF-MS	*Nicotiana attenuata*	I, root and shoots	*Manduca sexta*	Many	Control vs. induced/ Transcriptome-metabolome	Gulati et al. 2013
UHPLC-qTOF-MS, idMS/MS	*N. attenuata*	I, shoot	*Manduca sexta*	Many	Variation in induced response between different accessions	Li et al. 2015
GC-MS	Pepper (*Capsicum* spp.)	Shoot	*Frankliniella occidentalis*	Tocopherols, terpenes, sterols	Genotype comparison	Maharijaya et al. 2012
UHPLC-TOF-MS GC-MS	*Plantago lanceolata*	I, shoot	*Heliothis virescens*, *Myzus persicae*	Many	Multiple induction cross talk	Schweiger et al. 2014
GC-MS LC-MS	*P. lanceolata*	I, shoot	*Dysaphis cf. Plantaginea*, *Athalia circularis*, *Grammia incorupta*, *Heliothis virescens*	Many	Control vs. induced	Sutter and Müller 2011

Table 1 contd. ...

...Table 1 contd.

Metabolomics platform	**Plant species**	**Type of defense**	**Insect herbivore/ other organisms**	**Relevant metabolites found**	**Assessment**	**Reference**
GC-MS	*Populus tremula*	Shoot	Local insect herbivore community	Flavonoids, lipids, ascorbic acid	Population comparison/ Population genetics–metabolome –herbivore community	Bernhardsson et al. 2013
^{1}H NMR	Potato	C + I, Shoot	*Myzus persicae*	Glycoalkaloids, amino acids	GM vs. non-GM plants	Plischke et al. 2012
FT-ICR-MS	*Quercus robur*	C + I, shoot	*Tortrix viridana*	Flavones, tannins and terpenoids	Genotype comparison/ Transcriptome-metabolome	Kersten et al. 2013
^{1}H NMR	*Senecio* hybrids	C, Shoot	*Frankliniella occidentalis*	Chlorogenic acid, alkaloids	Genotype comparison	Leiss et al. 2009
Targeted LC-MS GC-MS	*Solanum galapagense*	C, shoot	Whitefly	Acyl sugars	Genotype comparison/QTL	Firdaus et al. 2013
UPLC-MS. MS/ MS	*Solanum nigrum*	I, shoot	*Noctuidae*	Many, e.g., phenolics	JA silenced plants vs. wild-type/ Gene expression-metabolome	Vandoorn et al. 2011
LC-qTOF-MS	13 species of Asteraceae	C + I, shoot	*Mamestra brassicae*	many, chlorogenic acid	Native vs. invasive species comparison	Macel et al. 2014

^{1}H-NMR	Thai Jasmin Rice (*Oryza sativa*)	C + I, shoot	Planthopper	amino acids	Cultivar comparison/ Control vs. induced	Uawisetwathana et al. 2015
GC-TOF-MS	Tomato (*Solanum lycopersicum*)	I, Root and shoot	*Manduca sexta*	Amino acids, sugars, organic acids, phenolics	Control vs. induced	Gomez et al. 2012
GC-MS	Tomato	I, Root and shoot	*Manduca sexta* *Helicoverpa zea*	Amino acids, phenolics	Control vs. induced	Steinbrenner et al. 2011
GC-MS	Tomato	I, shoot	Spider mite	HIPVs	Control vs. Induced/Gene expression	Kant 2004

than a decade of research by the same group on this plant-insect system (de Jager et al. 1996), the exact chemical compounds conferring defenses against thrips were largely unknown until metabolomics was applied. By comparing the metabolomes of resistant vs. non-resistant cultivars using ^{1}H-NMR, the common plant phenolic compound chlorogenic acid was identified as a major cause for thrips resistance. Chlorogenic acid was also found to play a role in thrips resistance in *Senecio* species and hybrids, which are Asteraceae like *Chrysanthemum* (Leiss et al. 2009b), but not in carrots (*Daucus carotus*) which belongs to the Umbelliferae (Leiss et al. 2013). Without using metabolomics, it would have been highly unlikely that an 'ordinary compound' such as chlorogenic acid would have been brought in relation to thrips resistance in Asterceae.

The majority of metabolomics studies on insect resistance thus far investigated induced resistance in the shoots by comparing the metabolomes of control (= undamaged) plants with herbivore-induced plants (Table 1). To elicit induced chemical defenses, either actual herbivores or mechanical wounding combined with the application of herbivore oral secretions were used. These studies yielded many compounds that were involved in both plant wounding responses and induced resistance to herbivores. Partly these were known metabolites, but almost always some hitherto unknown compounds were detected, showing the power of metabolomics as an exploratory tool for searching novel sources for (sustainable) insect resistance.

An untargeted metabolomics approach can also be applied to confirm that previously identified metabolites are indeed the only or most important compounds in induced responses of plants to insect feeding (Glauser et al. 2011). Only a few studies used the power of metabolomics to investigate which metabolites from the plants are taken up by the insects (Jansen et al. 2009; Glauser et al. 2011). The latter approach is especially relevant when taking a multitrophic perspective which includes natural enemies of the insect pest. Specialist herbivores that are metabolically adapted to their host plant may take up and sequester their host's defenses to protect themselves against their own enemies such as parasitoids and predators (Schoonhoven et al. 2005). This can affect the performance of higher trophic levels, and thus the effectiveness of biological control agents in integrated pest management strategies (IPM; Qiu et al. 2009). When breeding for sustainable resistance in an IPM context, it should also be considered whether or not the plant's chemical defenses are likely to be turned against itself via foodweb interactions.

Metabolomics studies on induced resistance have also been combined with transcriptome analyses (Table 1). Such systems biology approaches, as combinations of different—omics technologies are called (Weckwerth 2003), can assist the elucidation of gene functions and biochemical pathways

(Gulati et al. 2013). Not only shoot but also induced root metabolomes were analyzed (Gulati et al. 2013; Marti et al. 2013; Jansen et al. 2009; Gómez et al. 2012). Kutyniok and Müller (2012), for example, tested how induction by belowground feeding nematodes and aboveground herbivores, alone and in combination, altered root and shoot metabolomes. Using the genetic model species *Arabidopsis thaliana* their metabolomics analyses showed, amongst others, that glucosinolates play a role in the induced responses, which confirmed earlier targeted chemical analyses (van Dam et al. 2009). Recent technological developments with regards to the sensitivity of the detectors and computer processor speed, also allow for more detailed analyses of metabolites in the rhizosphere, for example, in root exudates or volatiles produced by bacteria (van Dam and Bouwmeester 2016).

Interestingly, metabolomics studies revealed that soil microbiota, e.g., rhizobacteria, can alter the shoot metabolome and subsequent shoot insect herbivory (Table 1; Badri et al. 2013; van de Mortel et al. 2012). Both studies used the same plant species (*A. thaliana*) but different analytical metabolomics platforms which yielded different relevant compound classes in the two studies, glucosinolates (LC-MS) and amino acids (GC-MS), respectively. Fungal endophytes can also affect plant chemistry. A metabolomics study on grass fungal endophytes revealed that the endophytes altered more metabolites in the plant than was previously known based on targeted alkaloid analyses (Rasmussen et al. 2009). The effects are often species-specific: in a multispecies comparison, the same fungal symbiont, an arbuscular mycorrhizal fungus, altered the metabolome of each plant species differentially (Schweiger et al. 2014). However, it was not clear how much of the variation was due to the fact that plant species have very specific metabolomes to begin with.

Applying Metabolomics in Plant Resistance Research

Metabolomics may be a very useful tool when searching for novel sources of plant resistance, especially in the first, exploratory phases of the project. Here we briefly describe and discuss the importance aspects that need to be considered at each step of the process (Fig. 1). For more detailed protocols, metabolomics workflows, or other resources needed in the course of the analytical processes, we refer to the appropriate sources.

To reduce natural background variation which is typical to plant metabolomics, ideally resistant and non-resistant Recombinant Inbred Lines (RILs) and/or Inbred Lines (IL) populations or genotypes are compared (see Li et al. 2015). When growing the plants, environmental factors should be controlled as much as possible. Both biotic (type of soil, soil biota) as well as abiotic (temperature, nutrients, light) environmental factors can strongly affect the plant's metabolome. For example,

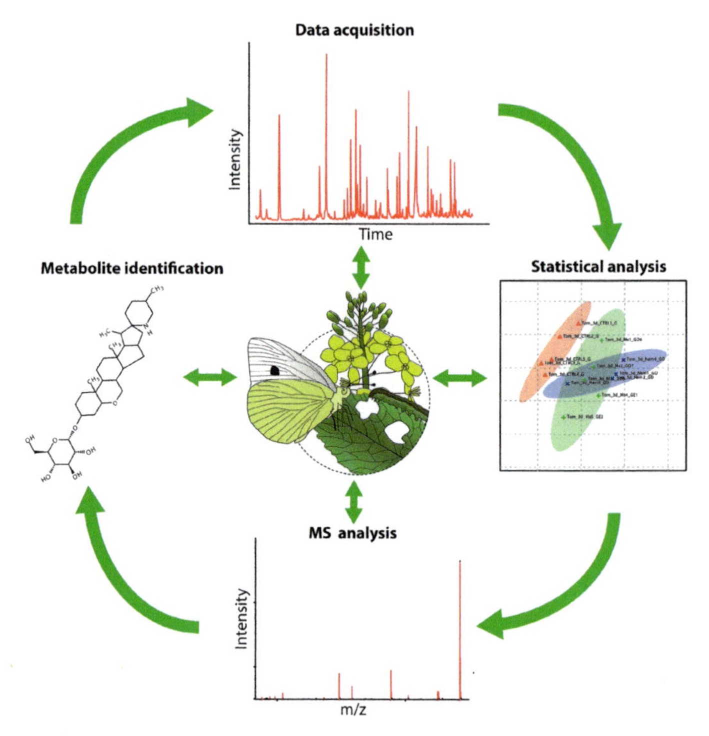

Figure 1. Flowchart of MS based plant metabolomic analysis. Genotypes or plants subjected to different treatments can be compared for their metabolomic profiles by using, for example, LC-MS. This generates large datasets of hundreds to thousands of mass signal intensities. The LC-MS results are statistically analyzed and metabolites of interests may be identified based on their MS fragmentation. Ideally, the effects of these metabolites should be tested on the target herbivore.

Original design: Alexander Weinhold and Onno Calf.

greenhouse-grown plants were found to differ in metabolomes depending on whether they were grown on shady spots or in the full sun (Jankanpaa et al. 2012). In addition the plant's metabolome changes during its ontogeny, especially when plants start to flower.

Bearing in mind the above, the following must be considered in order to obtain comparable samples for metabolomics analyses; (1) the type of plant material sampled should be as similar as possible among plants; young leaves have different chemical profiles than old leaves (van Dam et al. 1995), and therefore the age, or ontogenetic stage, of the plants sampled must be similar; (2) diurnal rhythms influence plant chemical profiles (Kim et al. 2011) and sampling should ideally be performed within in a narrow time window; when sampling takes places over multiple days, the time of day that sampling takes place should be similar; (3) to avoid the formation of break-down products, harvested plant material should be directly flash frozen in liquid nitrogen; (4) when studying induced responses, the timing of harvest after initial damage is crucial because the time-dependent nature of induced responses after herbivore attack (Mathur et al. 2011). In order to enhance the changes of finding relevant resistance factors, it would be advisable to sample at multiple time points after induction.

The analytical method chosen for metabolomics depends on the type of potentially interesting compounds (Macel et al. 2010). If the interest is in volatile organic compounds, Gas-Chromatography coupled with Mass Spectrometry is the most commonly used technique. For non-volatile compounds ^{1}H-NMR or Liquid-Chromatography coupled with Mass Spectrometry is more appropriate. There are a number of excellent metabolomics protocols published that cover the different analytical techniques, e.g., Kim, Choi and Verpoorte 2010 (NMR); Lisec et al. 2006 (GC-MS); De Vos et al. 2007 (LC-MS). For an example of a comparison of NMR and MS analyses see Porzel et al. (2014).

Each analytical approach has their own advantages and disadvantages. Most importantly, none of them covers the entire metabolome of a plant. This starts already with the selection of the extraction solvent, which will greatly determine the types of metabolites that will be extracted. Multiple extraction rounds with solvents differing in polarity will increase the metabolome coverage. For quality control reasons, the insertion of pooled samples, i.e., combined materials of different species or treatments, as well as an internal standard are necessary to check for technical errors. Randomization or block randomization of the order in which the samples are analyzed is advised when processing large sample sets, which is often the case in research on plant resistance screenings. Randomization of the sample order prohibits that the differences found between treatments or plant accessions are due to 'wear and tear' of the column or mass detectors, or other time bound effects.

After the chemical analyses are finished, the data need to be aligned and often normalized to be comparable. Mass peaks obtained on MS platforms, commonly are clustered into clusters of analytes/ion fragments that belong to the same metabolite based retention time and intensity profiles (Tikunov et al. 2005). These clusters are then referred to as 'features'. To further increase the quality of the dataset, the data can be filtered by removing those features that are only present in a few samples and are usually around the detection limit.

After data preprocessing, the dataset can be analyzed using multivariate techniques such as principal component analyses (PCA) and (orthogonal) partial-least-square discriminant analyses ((O)PLS-DA). While PCA is unbiased, PLS-DA uses a matrix with predefined, distinct classes, e.g., resistant vs. non-resistant. PLS-DA models are prone to overfitting and thus they may provide unreliable results. Therefore, validation of these statistical PLS-DA models is important. For further reading on metabolomics statistical data analyses see Jansen et al. (2010) and Westerhuis et al. (2008). The loading plots of PCA (in case of low number of features) or OPLS-DA (S-Plots) can be used to pinpoint relevant compounds that are for example different between resistant and non-resistant plant genotypes.

Eventually, the compounds relevant to insect resistance should be identified. This is not a trivial task, given that the majority of plant secondary metabolites is still unknown. However, with continuing technological advancements such as high resolution MS/MS approaches and novel search algorithms being developed, the identification of 'interesting unknowns' is rapidly becoming more feasible. There are several metabolomics databases available online, which are particularly helpful for GC-MS and ^{1}H-NMR. Links to these databases can be found at the website of the metabolomics society: http://metabolomicssociety.org/resources/metabolomics-databases. Unfortunately, these databases are as yet less useful for LC-MS. Often, metabolomics research labs create their own library that is fine-tuned to their specific equipment and procedures. Ideally, unknown metabolites correlating with resistance are isolated and identified by ^{1}H-NMR and/or ^{13}C-NMR spectroscopy (Prince and Pohnert 2010). The latter is often challenging as up to 1 mg of the pure compound may be needed for proper NMR analysis.

Even for known metabolites, their exact function in plant-insect interactions is often still unknown. If resistant genotypes contain compound X that is lacking in non-resistant genotypes, it seems likely that compound X is involved in the insect resistance of the plants. However, this relation is correlative, thus the function of compound X in resistance needs further investigation. This can be done by bioassays where the effect of the isolated compound X on insect feeding is explicitly tested (e.g., Leiss et al. 2009). In case the biosynthetic pathway is known, gene silencing (RNAi experiments, CRISPR/Cas9) may be used to produce plants without compound X (e.g., Li et al. 2015). The metabolomics analyses itself does not show if compound X is active alone or in combination with other metabolites, nor does it reveal which biosynthetic pathway is used for the production of the compound (but see Li et al. 2015 for a novel molecular network construction approach).

Linking Metabolomics with Genetic Analyses

Knowledge on which compounds—or more likely—groups of compounds cause insect resistance in breeding stock will not suffice, as it is not known if and how the traits would be inherited. Breeders often apply Quantitative Trait Loci (QTL) analyses to identify specific desirable plant traits in the genome of their crops. The genetic control of the metabolome can be investigated by coupling metabolomics data to genetic analyses in for example mQTL (metabolic Quantitative Trait Loci) for constitutive defenses. Similarly, genetic maps obtained through genome wide association mapping (GWAS) can be linked to high resolution metabolomics profiling. This can reveal genomic regions involved in the production of particular metabolites, as well as insect resistance, and can also further elucidate biosynthetic pathways. In rice, for example, this has been applied to pinpoint genes involved in flavonoid production (Chen et al. 2014). The same study also reconstructed metabolic pathways based on the mQTLs. The few studies thus far that performed GWAS in combination with widely targeted metabolomics were performed on rice (Chen et al. 2014; Matsuda et al. 2015; Gong et al. 2013), tomato (Alseekh et al. 2015), maize (Riedelsheimer et al. 2012), and *Arabidopsis* (Keurentjes et al. 2006; Rowe et al. 2008; Joseph et al. 2014). Matsuda et al. (2015) showed that levels of a large number of metabolites in rice are tightly associated with a relatively small number of strong QTLs. In addition, a combined mQTL and expression QTL (eQTL) study on *Brassica rapa* revealed complex genetic linkages for regulation of several types of metabolites (Del Carpio et al. 2014). The latter study also shows that the genetic regulation of many compounds is genetically linked. This makes it difficult to positively select for a single chemical trait in plant breeding without 'dragging along' other compounds. In principle, this does not have to be a problem, unless there are negative effects on other qualities of the crop, such a flavor or yield. Other studies used a more narrow chemical approach (for example glucosinolate or fatty acid analyses) and coupled this with genomics and transcriptomics (e.g., eQTL) to try to understand the genetic regulation of specific metabolites (e.g., Chan et al. 2010; Jensen et al. 2014; Basnet et al. 2016).

The genetics of herbivore-induced resistance can be studied by coupling transcriptomes with metabolomes. Transcriptome studies linked to metabolomics in model systems has led to the identification of novel gene functions in the terpenoid and glucosinolate biosynthesis, in maize and *Arabidopsis* respectively (Tzin et al. 2015; Hirai et al. 2007). Next generation RNA-sequencing (RNAseq) has greatly advanced genomic studies of non-model species and does not require the production of inbred lines (Martin et al. 2013). As a consequence, the combination of RNAseq and metabolomics is more readily applied to (long lived) non-model systems as well. RNAseq combined with metabolomics in *Quercus robur* (penduncula te oak) revealed several hundreds of genes that were differentially expressed, either constitutively or after induction, between susceptible and resistant oaks. Groups of metabolites that were linked to this differential gene expression were flavones, tannins and terpenoids (Kersten et al. 2013). Interestingly, herbivore resistant oaks had higher constitutive expression of defenses while the susceptible oak genotypes were more inducible. A study on Poplar linked untargeted metabolomics with the expression of particular known defense genes and local insect herbivore communities of several populations (Bernhardsson et al. 2013). The geographic structure of all three analyses tightly matched, indicating an evolutionary arms-race between the tree and its herbivores.

Outlook

Metabolomics offers a great potential to gain new insights into the mechanisms underlying plant resistance to insect herbivores. This is mainly because it allows for the discovery of novel compounds as well as combinations of compounds conferring resistance to pest insects. The identification of the relevant metabolites and the assessment of their exact function is still a time-consuming process, and is likely the bottleneck for application of metabolomics in breeding programs. However, with the fast development of novel bioinformatics approaches to search multiple databases, the identification of compounds in a metabolomics profile identification is becoming increasingly easier (Neumann and Böcker 2010; Hufsky and Böcker 2016). As outlined before, a systems biology approach, which links several -omics techniques at different levels, is a promising approach for understanding mechanisms of plant resistance and facilitates its implementation in crop breeding. Also here, advancements in bioinformatics are crucial to interconnect the large amount of data that is generated at each -omics level, e.g., genomics, transcriptomics and metabolomics. Currently, bioinformatics tools are developed for each level separately, and relatively few have endeavored to combine datasets from these different levels of organization. Pathway analyses linking different -omics techniques could help overcome this caveat (Kaever et al. 2015). When using genetically modified crops, or novel, more readily accepted tools for genome editing such as CRISPR/CAS9, metabolomics can also be used to test for unforeseen changes in the metabolome of modified plants (Plischke et al. 2012; Kuiper et al. 2001). However, due to the lack of knowledge on the bioactivity of many of the compounds, metabolomics is not a tool that can used for Environmental Risk Assessments (ERA) yet (Hall and de Maagd 2014).

mQTLs are an example of how combining metabolomics with genetic analyses can be very useful for metabolite-assisted breeding and marker-assisted selection of crops. One crucial limitation though, is the heritability of the metabolome. Some studies showed relatively low heritabilities for at least part of the metabolome (Laurentin et al. 2008). Often there are also stochastic effects on gene expression in relation to plant defenses. It has been suggested that part of this stochastic variation is under genetic control. In natural environments random variation in traits can be adaptive. For plant breeders who target for a constant product quality, uncontrolled variability is an unwanted trait. More research has to go into defining the regulatory and genetic mechanisms underlying this variation before it can be controlled.

Metabolomics has thus far been mostly used in screening crops for food quality (e.g., Basnet et al. 2016; Ghan et al. 2015; Li et al. 2013). However, our review reveals several studies that used metabolomics to screen for (novel) chemical mechanisms of insect resistance in crops such as pepper, citrus or tomato (Table 1). For a large part this may be due to the fact that several effective synthetic pesticides, such as neonicotenoids, have been banned. Breeders are therefore forced to search for natural insect resistance in their stock collections. When properly combined with marker-assisted breeding and trait discovery, metabolomics is an excellent tool to contribute to the rapid development of more sustainable resistance in crop plants.

Acknowledgments

Nicole M. van Dam gratefully acknowledges the support of the German Centre for Integrative Biodiversity Research (iDiv) Halle-Jena-Leipzig funded by the German Research Foundation (FZT 118) and the Dutch Technology Foundation STW for funding the project "Identifying new sources of thrips resistance in *Capsicum*" (Project number 13552) within the Program funding to "Green defences Against Pests" (GAP).

References

Agut, B., J. Gamir, J.A. Jacques et al. 2015. Tetranychus urticae-triggered responses promote genotype-dependent conspecific repellence or attractiveness in citrus. New Phytol. 207: 790–804.

Alseekh, S., T. Tohge, R. Wendenberg et al. 2015. Identification and mode of inheritance of quantitative trait loci for secondary metabolite abundance in tomato. The Plant Cell. 27: 485–512.

Badri, D.V., G. Zolla, M.D. Bakker et al. 2013. Potential impact of soil microbiomes on the leaf metabolome and on herbivore feeding behavior. The New Phytol. 198: 264–73.

Basnet, R.K., D.P. Del Carpio, D. Xiao et al. 2016. A systems genetics approach identifies gene regulatory networks associated with fatty acid composition in *Brassica rapa* seed. Plant Physiol. 170: 568–585.

Beekwilder, J., W. van Leeuwen, N.M. Van Dam et al. 2008. The impact of the absence of aliphatic glucosinolates on insect herbivory in Arabidopsis. PLoS ONE. 3: e2068.

Bernhardsson, C., K.M. Robinson, I.N. Abreu et al. 2013. Geographic structure in metabolome and herbivore community co-occurs with genetic structure in plant defence genes. Ecol Lett. 16: 791–798.

Cajka, T. and O. Fiehn. 2016. Toward merging untargeted and targeted methods in mass spectrometry-based metabolomics and lipidomics. Anal Chem. 88: 524–545.

Chan, E.K.F., H.C. Rowe and D.J. Kliebenstein. 2010. Understanding the evolution of defense metabolites in Arabidopsis thaliana using genome-wide association mapping. Genetics. 185: 991–1007.

Chen, W., Y. Gao, X. Weibo et al. 2014. Genome-wide association analyses provide genetic and biochemical insights into natural variation in rice metabolism. Nat Genet. 46: 714–21.

de Jager, C.M., R.P.T. Butot, E. Vander Meijden et al. 1996. The role of primary and secondary metabolites in chrysanthemum resistance to Frankliniella occidentalis. J Chem Ecol. 22: 1987–1999.

De Vos, R.C.H., S. Moco, A. Lommen et al. 2007. Untargeted large-scale plant metabolomics using liquid chromatography coupled to mass spectrometry. Nat Prot. 2: 778–791.

Del Carpio, D.P., R.k. Basnet, D. Arends et al. 2014. Regulatory network of secondary metabolism in Brassica rapa: insight into the glucosinolate pathway. PloS One. 9: e107123.

Dhami, M.K., R. Gardner-Gee, J.V. Houtte et al. 2011. Species-specific chemical signatures in scale insect honeydew. J Chem Ecol. 37: 1231–1241.

Erb, M., J. Ton, J. Degenhardt et al. 2008. Interactions between arthropod-induced aboveground and belowground defenses in plants. Plant Phys. 146: 867–874.

Fatouros, N.E., D. Lucas Barbosa, B.T. Weldegergis et al. 2012. Plant volatiles induced by herbivore egg deposition affect insects of different trophic levels. PLoS ONE. 7: e43607.

Fiehn, O., J. Kopka, P. Dormann et al. 2000. Metabolite profiling for plant functional genomics. Nat Biotechnol. 18: 1157–1161.

Firdaus, S., A.W. van Heusden, N. Hidayati et al. 2013. Identification and QTL mapping of whitefly resistance components in *Solanum galapagense*. Theor Appl Genet. 126: 1487–1501.

Gassmann, A.J., J.L. Petzold-Maxwell, R.S. Keweshan et al. 2011. Field-evolved resistance to Bt maize by western corn rootworm. PLoS ONE. 6: e22629.

Ghan, R., S.C. van Sluyter, U. Hochberg et al. 2015. Five omic technologies are concordant in differentiating the biochemical characteristics of the berries of five grapevine (*Vitis vinifera* L.) cultivars. BMC Genomics. 16: 946.

Glauser, G., G. Marti, N. Villard et al. 2011. Induction and detoxification of maize 1,4-benzoxazin-3-ones by insect herbivores. Plant J. 68: 901–911.

Gómez, S., A.D. Steinbrenner, S. Ossorio et al. 2012. From shoots to roots: Transport and metabolic changes in tomato after simulated feeding by a specialist lepidopteran. Entomol Exper Appl. 144: 101–111.

Gong, L., W. Chen, S. Gao et al. 2013. Genetic analysis of the metabolome exemplified using a rice population. Proc Natl Acad Sci USA. 110: 20320–20325.

Gulati, J., S.G. Kim, I.T. Baldwin et al. 2013. Deciphering herbivory-induced gene-to-metabolite dynamics in *Nicotiana attenuata* tissues using a multifactorial approach. Plant Physiol. 162: 1042–1059.

Hall, R.D. and R.A. de Maagd. 2014. Plant metabolomics is not ripe for environmental risk assessment. Trends Biotechnol. 32: 391–392.

Hirai, M.Y., K. Sugiyami, Y. Sawada et al. 2007. Omics-based identification of Arabidopsis Myb transcription factors regulating aliphatic glucosinolate biosynthesis. Proc Natl Acad Sci USA. 140: 6478–6483.

Hol, W.H., W. de Boehr, A.J. Termorshuzien et al. 2010. Reduction of rare soil microbes modifies plant—herbivore interactions. Ecol Lett. 13: 292–301.

Hufsky, F. and S. Böcker. 2016. Mining molecular structure databases: Identification of small molecules based on fragmentation mass spectrometry data. Mass Spect Rev. 9999: XX–XX.

Jankanpaa, H.J., Y. Mishra, W.P. Schroeder et al. 2012. Metabolic profiling reveals metabolic shifts in Arabidopsis plants grown under different light conditions. Plant Cell Environ. 35: 1824–1836.

Jansen, J.J., J.W. Allwood, E. Marsden-Edwards et al. 2009. Metabolomic analysis of the interaction between plants and herbivores. Metabol. 5: 150–161.

Jansen, J.J., S. Smit, H.C. Hoefsloot et al. 2010. The photographer and the greenhouse: How to analyse plant metabolomics data. Phytochem Anal. 21: 48–60.

Joseph, B., S. Atwell, J.A. Corwin et al. 2014. Meta-analysis of metabolome QTLs in Arabidopsis: trying to estimate the network size controlling genetic variation of the metabolome. Front Plant Sci. 5: 461.

Joseph, B., J.A. Corwin and D.J. Kliebenstein. 2015. Genetic variation in the nuclear and organellar genomes modulates stochastic variation in the metabolome, growth, and defense. PLoS Genetics. 11(1): e1004779.

Kaever, A., M. Landesfeind, K. Feussner et al. 2015. MarVis-Pathway: integrative and exploratory pathway analysis of non-targeted metabolomics data. Metabol. 11: 764–777.

Kant, M.R. 2004. Differential timing of spider mite-induced direct and indirect defenses in tomato plants. Plant Physiol. 135: 483–495.

Karban, R. and I.T. Baldwin. 1997. Induced Responses to Herbivory. University of Chicago Press, Chicago, U.S.A.

Kersten, B., A. Ghirardo, J.P. Schnitzler et al. 2013. Integrated transcriptomics and metabolomics decipher differences in the resistance of pedunculate oak to the herbivore *Tortrix viridana* L. BMC Genomics. 14: 737.

Keurentjes, J.J.B., J. Fu, C.H.R. de Vos et al. 2006. The genetics of plant metabolism. Nat Genet. 38: 842–849.

Kuiper, H.A., G.A. Kleter, H.P. Noteborn et al. 2001. Assessment of the food safety issues related to genetically modified foods. The Plant J. 27: 503–528.

Kutyniok, M. and C. Mueller. 2012. Crosstalk between above- and belowground herbivores is mediated by minute metabolic responses of the host *Arabidopsis thaliana*. J Exptl Bot. 63: 6199–6210.

Kuzina, V., C.T. Eckstrom, S.B. Andersen et al. 2009. Identification of defense compounds in *Barbarea vulgaris* against the herbivore *Phyllotreta nemorum* by an ecometabolomic approach. Plant Physiol. 151: 1977–1990.

Kuzina, V., J.K. Nielsen, J.M. Augustin et al. 2011. *Barbarea vulgaris* linkage map and quantitative trait loci for saponins, glucosinolates, hairiness and resistance to the herbivore *Phyllotreta nemorum*. Phytochem. 72: 188–198.

Laurentin, H., A. Ratzinger and P. Karlovsky. 2008. Relationship between metabolic and genomic diversity in sesame (*Sesamum indicum* L.). BMC Genomics. 9: 250.

Leiss, K.A., F. Maltese, Y.H. Choi et al. 2009a. Identification of chlorogenic acid as a resistance factor for thrips in chrysanthemum. Plant Physiol. 150: 1567–1575.

Leiss, K.A., Y.H. Choi, I.B. Abdel Farid et al. 2009b. NMR metabolomics of thrips (*Frankliniella occidentalis*) resistance in senecio hybrids. J Chem Ecol. 35: 219–229.

Leiss, K.A., G. Cristofori, R. van Steenis et al. 2013. An eco-metabolomic study of host plant resistance to Western flower thrips in cultivated, biofortified and wild carrots. Phytochem. 93: 63–70.

Li, D., I.T. Baldwin and E. Gaquerel. 2015. Navigating natural variation in herbivory-induced secondary metabolism in coyote tobacco populations using MS/MS structural analysis. Proc Natl Acad Sci USA. 112: E4147–4155.

Li, W., C.-J. Ruan, J.A. Teixeira et al. 2013. NMR metabolomics of berry quality in sea buckthorn (*Hippophae* L.). Mol Breed. 31: 57–67.

Lisec, J., N. Schauer, J. Kopka et al. 2006. Gas chromatography mass spectrometry-based metabolite profiling in plants. Nat Prot. 1: 387–396.

Macel, M., N.M. van Dam and J.J.B. Keurentjes. 2010. Metabolomics: the chemistry between ecology and genetics. Mol Ecol Resour. 10: 583–593.

Macel, M., R.C.H. de Vos, J.J. Jansen et al. 2014. Novel chemistry of invasive plants: Exotic species have more unique metabolomic profiles than native congeners. Ecol Evol. 4: 2777–2786.

Maharijaya, A., B. Vosman, F. Verstappen et al. 2012. Resistance factors in pepper inhibit larval development of thrips (*Frankliniella occidentalis*). Entomol Exptl Applic. 145: 62–71.

Marti, G., M. Erb, J. Boccard et al. 2013. Metabolomics reveals herbivore-induced metabolites of resistance and susceptibility in maize leaves and roots. Plant Cell Environ. 36: 621–639.

Martin, L.B.B., Z. Fei, J.J. Giovanonni et al. 2013. Catalyzing plant science research with RNA-seq. Front Plant Sci. 4: 66.

Mathur, V., S. Ganta, C.E. Raaijmakers et al. 2011. Temporal dynamics of herbivore-induced responses in Brassica juncea and their effect on generalist and specialist herbivores. Entomol Exptlis Applic. 139: 215–225.

Matsuda, F., R. Nakabayashi, Z. Yang et al. 2015. Metabolome-genome-wide association study dissects genetic architecture for generating natural variation in rice secondary metabolism. Plant J. 81: 13–23.

Miyagi, A., M.K. Yamada, M. Uchimiya et al. 2013. Metabolome analysis of food-chain between plants and insects. Metabol. 9: 1254–1261.

Mohanta, T.K., A. Occipinti, Z.S. Attsabha et al. 2012. *Ginkgo biloba* responds to herbivory by activating early signaling and direct defenses. PLoS ONE. 7: e32822.

Neumann, S. and S. Böcker. 2010. Computational mass spectrometry for metabolomics: Identification of metabolites and small molecules. Anal Bioanal Chem. 398: 2779–2788.

Nguyen, D., N. D'Agostin, T.O.G. Tytgat et al. 2016. Drought and flooding have distinct effects on herbivore-induced responses and resistance in Solanum dulcamara. Plant Cell Env:n/a-n/a. doi: 10.1111/pce.12708.

Ossipov, V., T. Klemola, K. Ruohomäki et al. 2014. Effects of three years` increase in density of the geometrid Epirrita autumnata on the change in metabolome of mountain birch trees (Betula pubescens ssp. czerepanovii). Chemoecol. 24: 201–214.

Plischke, A., Y.H. Choi, P.M. Brakefield et al. 2012. Metabolomic plasticity in GM and non-GM potato leaves in response to aphid herbivory and virus infection. J Agric Food Chem. 60: 1488–1493.

Porzel, A., M.A. Farag and J. Mulbradt. 2014. Metabolite profiling and fingerprinting of *Hypericum* species: A comparison of MS and NMR metabolomics. Metabol. 10: 574–588.

Prince, E.K. and G. Pohnert. 2010. Searching for signals in the noise: metabolomics in chemical ecology. Anal Bioanal Chem. 396: 193–197.

Qiu, B.L., J.A. Harvey, C.E. Raaijmakers et al. 2009. Nonlinear effects of plant root and shoot jasmonic acid application on the performance of Pieris brassicae and its parasitoid *Cotesia glomerata*. Functional Ecol. 23: 496–505.

Rasmussen, S., A.J. Parsons and J.A. Newman. 2009. Metabolomics analysis of the Lolium perenne-Neotyphodium lolii symbiosis: More than just alkaloids? Phytochem Rev. 8: 535–550.

Riedelsheimer, C., J. Lisec, A. Czedic Eyesenberg et al. 2012. Genome-wide association mapping of leaf metabolic profiles for dissecting complex traits in maize. Proc Natl Acad Sci USA. 109: 8872–8877.

Rowe, H.C., B.J. Hansen, B.A. Halkier et al. 2008. Biochemical networks and epistasis shape the *Arabidopsis thaliana* metabolome. The Plant Cell. 20: 1199–1216.

Schoonhoven, L.M., J.J.A. Van Loon and M. Dicke. 2005. Insect-Plant Biology. Oxford University Press, Oxford, UK.

Schweiger, R., M.C. Baier, M. Persicke et al. 2014. High specificity in plant leaf metabolic responses to arbuscular mycorrhiza. Nat Comm. 5: 3886.

Shepherd, L.V.T., C.A. Hackett, C.J. Alexander et al. 2015. Modifying glycoalkaloid content in transgenic potato—Metabolome impacts. Food Chem. 187: 437–443.

Simo, C., C. Ibanez, A. Valdes et al. 2014. Metabolomics of genetically modified crops. Intl J Mol Sci. 15: 18941–18966.

Soler, R., J.A. Harvey, A.F.D. Kamp et al. 2007. Root herbivores influence the behaviour of an aboveground parasitoid through changes in plant-volatile signals. Oikos. 116: 367–376.

Soltis, N.E. and D.J. Kliebenstein. 2015. Natural variation of plant metabolism: genetic mechanisms, interpretive caveats, evolutionary and mechanistic insights. Plant Physiol. 169: 1456–1478.

Steinbrenner, A.D., S. Gomez, S. Osorio et al. 2011. Herbivore-induced changes in tomato (Solanum lycopersicum) primary metabolism: A whole plant perspective. J Chem Ecol. 37: 1294–1303.

Sutter, R. and C. Müller. 2011. Mining for treatment-specific and general changes in target compounds and metabolic fingerprints in response to herbivory and phytohormones in *Plantago lanceolata*. New Phytol. 191: 1069–1082.

Tikunov, Y., A. Lommen, C.H. de Vos et al. 2005. A novel approach for nontargeted data analysis for metabolomics. Large-scale profiling of tomato fruit volatiles. Plant Physiol. 139: 1125–1137.

Tzin, V., N. Fernandez-Pozo, A. Richter et al. 2015. Dynamic maize responses to aphid feeding are revealed by a time series of transcriptomic and metabolomic assays. Plant Physiol. 169: 1727–1743.

Uawisetwathana, U., S.F. Graham, W. Kamolsukyunyong et al. 2015. Quantitative 1H NMR metabolome profiling of Thai Jasmine rice (*Oryza sativa*) reveals primary metabolic response during brown planthopper infestation. Metabol. 11: 1640–1655.

van Dam, N.M., L.W.M. Vuister, C. Bergshoeff et al. 1995. The "Raison-D"etre' of pyrrolizidine alkaloids in Cynoglossum officinale: deterrent effects against generalist herbivores. J Chem Ecol. 21: 507–523.

van Dam, N.M. and H.J. Bouwmeester. 2016. Metabolomics in the Rhizosphere: Tapping into belowground chemical communication. Trends Plant Sci. 21: 256–265.

van de Mortel, J.E., R.C. de Vos, E. Dekkers et al. 2012. Metabolic and transcriptomic changes induced in Arabidopsis by the rhizobacterium *Pseudomonas fluorescens* SS101. Plant Physiol. 160: 2173–2188.

Vandoorn, A., G. Bonaventure, I. Rogachev et al. 2011. JA-Ile signalling in *Solanum nigrum* is not required for defence responses in nature. Plant Cell Environ. 34: 2159–2171.

Weckwerth, W. 2003. Metabolomics in systems biology. Ann Rev Plant Biol. 54: 669–689.

Weeks, D.P., M.H. Spalding and B. Yang. 2016. Use of designer nucleases for targeted gene and genome editing in plants. Plant Biotechnol J. 14: 483–495.

Westerhuis, J.A., H.C.J. Hoefsloot, S. Smit et al. 2008. Assessment of PLSDA cross validation. Metabol. 4: 81–89.

Widarto, H.T., E. Van Der Meijden, A.W. Lefeber et al. 2006. Metabolomic differentiation of Brassica rapa following herbivory by different insect instars using two-dimensional nuclear magnetic resonance spectroscopy. J Chem Ecol. 32: 2417–2428.

CHAPTER 8

RNAi and microRNA Technologies to Combat Plant Insect Pests

Vemanna S. Ramu, *K.C. Babitha* and *Kirankumar S. Mysore**

INTRODUCTION

Insects infect a wide range of economically important crops across the globe and cost billions of dollars for agriculture. The worldwide damage caused by pests (insects, plant pathogens and weeds) accounts for nearly an average of 37% of crop loss per year (Pimental 2005) and varies from crop to crop. For example, 50% in wheat, more than 80% in cotton, 26–29% in soybean, 31% in maize, 37% in rice and 40% in potatoes has been reported (Oerke 2006). The use of pesticides is a common approach around the world to control pests, but they are associated with significant hazards to the environment and human health. In addition, pesticides significantly increase the input cost of agriculture. Even though integrated pest management practices have been employed to control some pests it is not always effective. Therefore, there is a need to develop genetic control methods to reduce pesticide usage. Several approaches exist to develop insect/pest resistant varieties. Classical breeding methods have been used to develop various insect/pest resistant cultivars but it is time consuming and tedious as complexity increases with other added traits. Biotechnological approaches involve the expression of *Bacillus thuringiensis*

Noble Research Institute, LLC, 2510, Sam Noble Parkway, Ardmore, Oklahoma, USA 73401.
Email: rsvemanna@noble.org; kcbabitha@noble.org
* Corresponding author: ksmysore@noble.org

(Bt) insecticidal proteins (Cry toxins) in plants. Even though this approach is very effective the main drawback of this approach is that a group of Cry protein is specific to a particular family of insects. For example, the Cry1a group of toxins target Lepidoptera (butterflies) family, and Cry3 group is effective against beetles. More than 100 different *Bt* toxins have been identified in diverse strains of *B. thuringiensis* (Sanahuja et al. 2011). In addition to this drawback the effectiveness of *Bt* toxins is threatened by the development of *Bt* resistance in some insect species such as *Ostrinia nubilalis* (Lepidoptera; Pyralidae) and *Heliothis virescens* (Lepidoptera: Noctuidae) (Ferre and Van Rie 2002; Baum et al. 2007). Furthermore, there is increased resistance from environmentalists against this technology (Kumar et al. 2008; Benbrook 2012). This necessitates the development of other alternatives for pest control. Several natural insecticidal compounds are available as alternatives to *Bt* toxin. Amongst them, chitinase, lectins, alpha-amylase inhibitors, proteinase inhibitors, and cystatin proteins have shown potential in controlling pests (Gatehouse et al. 2011). Genetically modified plants expressing these defense responsive proteins are still in early stages of development.

Another alternative approach for insect control is the use of RNA interference (RNAi) technology. It is referred as post-transcriptional gene silencing (PTGS) in plants (vander Krol et al. 1990), quelling in fungi (Cogoni et al. 1996) and RNA interference in animals (Fire et al. 1998). Globally, RNAi is being used as one of the important tools in functional genomics studies to suppress a gene of interest and thereby link a phenotype to gene function. In animal system, RNAi has been used to control cancer and viral diseases (Huvenne and Smagghe 2010). In plants, RNAi technology has been used to create valuable crop traits like resistance to viruses, bacteria and nematodes. In addition to commercial applications, RNAi provides an avenue for functional genomics studies in animals, insects, plants and fungi (Belles 2010). RNAi has enormous potential for applied entomology (Price and Gatehouse 2008; Xue et al. 2012). RNAi-mediated gene silencing has been proven to control different orders of insects such as *Coleoptera, Hemiptera, Hymenoptera, Lepidoptera and Orthoptera* by suppressing essential genes leading to reduced fitness and/or mortality (Gatehouse and Price 2011; Zhang et al. 2013).

RNAi is highly conserved in eukaryotic organisms (Fire 2007). It is considered as a type of defence mechanism especially against viruses (Terenius et al. 2011). Four different types of RNAi have been described including short interfering RNAs (siRNAs), piwi-interacting RNAs (piRNAs), endogenous siRNAs (endo-siRNAs or esiRNAs) and microRNAs (miRNAs) (Terenius et al. 2011).

Silencing of several key pathogen mRNA by plant-expressed RNAi constructs showed resistance against important pathogens such as nematodes, fungi or aphids. This phenomenon is called as Host-Induced Gene Silencing (HIGS) and showed potential application in plant protection without the need of chemical disease control mechanisms. Similar strategies were successfully employed in insects by targeting key insect mRNAs (Baum et al. 2007; Mao et al. 2007, 2013; Gatehouse and Price 2011; Zhang et al. 2013).

> Ag-bio giant Monsanto in early 2015 announced that RNAi-based pesticide for the control of the Colorado potato beetle (CPB), a pest that attacks potatoes and other solanaceous plants, has advanced out of the discovery stage and will now undergo formal product development.
> (https://www.genomeweb.com/mirnarnai/monsanto-advances-topical-rnai-pesticide-against-colorado-potato-beetle)

The GM corn plants developed against western corn root worm (*Diabrotica virgifera*), by inhibiting the translation of *vacuolar H+-ATPase subunit A (v-ATPase A)* in the insect resulted in increased pest mortality and larval stunting and less root damage in corn (Baum et al. 2007). The rice transgenic plants showing suppression of several genes against *Nilaparvata lugens* (Brown planthopper), a major rice pest, resulted in reduced growth of insect (Zha et al. 2011). In another study for pest management, Mao and colleagues (2011) developed transgenic cotton plants to produce double-stranded RNA (dsRNA) against cotton bollworms (*Helicoverpa armigera*) by silencing the expression of *P450* gene *CYP6AE14*. Furthermore, by priming the antiviral RNAi response with innocuous viral sequences, beneficial insect species, such as honey bee (*Apis mellifera*) and silkworm (*Bombyx mori*), can be protected from highly pathogenic viral infections (Xue et al. 2012). These studies show that the RNAi-based GM crops that are lethal to pests or that deleteriously affect interactions of the pests with other organisms (including the crop) has the potential for limiting the impact of pests on crops. However, increasing research evidences suggest that there is a high level of complexity of gene regulation than originally anticipated with regards to uptake and delivery of DsRNA to the targeted site (Bartel 2009). RNAi based genetically modified (GM) plants may revolutionize the insect control methods than traditionally applied insecticides and *Bt* expressing GM plants. Several efforts are underway to develop RNAi plants but are still in their infant stages of laboratory tests.

Mechanism of RNAi

RNAi refers to PTGS by small non-coding RNAs, predominantly by the cleavage of a target mRNA in a sequence-specific manner (Fire et al. 1998).

Two main classes of small non-coding RNAs (sRNAs) are microRNAs (miRNAs) and short interfering RNAs (siRNAs). The biogenesis of miRNAs and siRNA are different but both have common elements. The complex double stranded RNAs (dsRNAs) serve as precursor for both miRNA and siRNA biogenesis by recognition of ribonuclease III enzyme Dicer and association with an Argonaut family protein (AGO) (Ketting 2011). The dsRNA is cleaved by the RNase III Dicer into 20–25 bp fragments with a two base overhang at the 3′ end. The 20–25 bp RNAs generated by Dicer can either be miRNAs or siRNAs depending on the precursor (Ghildiyal and Zamore 2009; Matranga and Zamore 2007; Asgari 2013). These fragments are incorporated into the multi-protein RNA-induced silencing complex (RISC), where one strand (the "passenger" strand) is eliminated and the other "guide" strand is retained. The catalytic component of RISC is the RNase H-like domain of an AGO protein, which cleaves single stranded RNA molecules having sequence complementary to the guide RNA. The RISC directed by the RNA guide strand, locates mRNAs containing specific nucleotide sequences complementary to the guide, binds to these sequences and cleaves the transcripts (Fig. 1) (Siomi and Siomi 2009; Winter et al. 2009).

The experimental use of RNAi exploits the siRNA pathway, specifically the capacity of cells to degrade a single-stranded RNA (ssRNA) (including mRNAs) with sequence identity to the administered dsRNA molecules (Roether and Meister 2011) (Fig. 1). The first step in the siRNA pathway is the recognition of the dsRNA by a complex consisting of a transactivation response RNA binding protein (TRBP) and the RNase III enzyme (Dicer-2), which cuts the molecule into small 21–23 bp siRNAs. The guide strand of the siRNA then enters the RISC bound to AGO followed by binding of the complex to complementary sequences of RNA targets leading to degradation of the target (Roether and Meister 2011). The most interesting aspect of RNAi is dsRNA rather than single-stranded antisense RNA and is highly sequence specific and remarkably potent (only a few dsRNA molecules per cell are required for effective interference). The interfering activity (presumably the dsRNA) can cause interference in cells and tissues far away from the site of siRNA synthesis.

In insects, RNAi-mediated silencing is transmitted widely throughout the organism therefore RNAi is described as systemic (Whangbo and Hunter 2008; Huvenne and Smagghe 2010). The processes determining the efficacy of RNAi are cellular uptake of the RNAi molecule, production of secondary dsRNA molecules in the cell and transfer of dsRNA molecules to other cells. In principle, the success of RNAi can be predicted from the functional level of these activities in the organism of interest and strategies that increase these activities might enhance RNAi-mediated knockdown of the target gene expression.

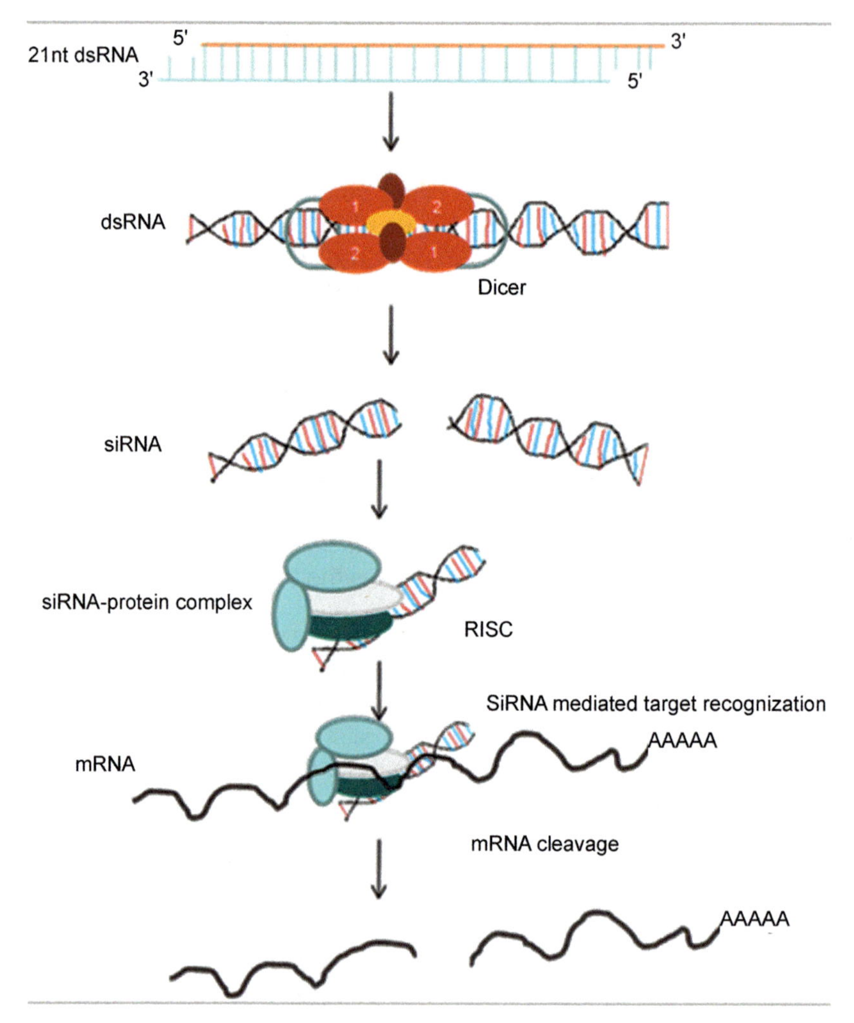

Figure 1. A schematic representation of the production of dsRNA and silencing the target gene.

RNAi in Insects

RNAi technology is also used in the field of entomology to study the function, regulation and expression of gene cascades, predominantly in *Drosophila melanogaster*, *Tribolium castaneum* and *Bombyx mori* (Fig. 2) (Huvene et al. 2010). The efficacy of RNAi varies across insect taxa, among genes, and with mode of delivery (Terenius et al. 2011). Early RNAi experiments in insects involved direct injection of dsRNA into the organism. However, the direct injection method is not suitable for large

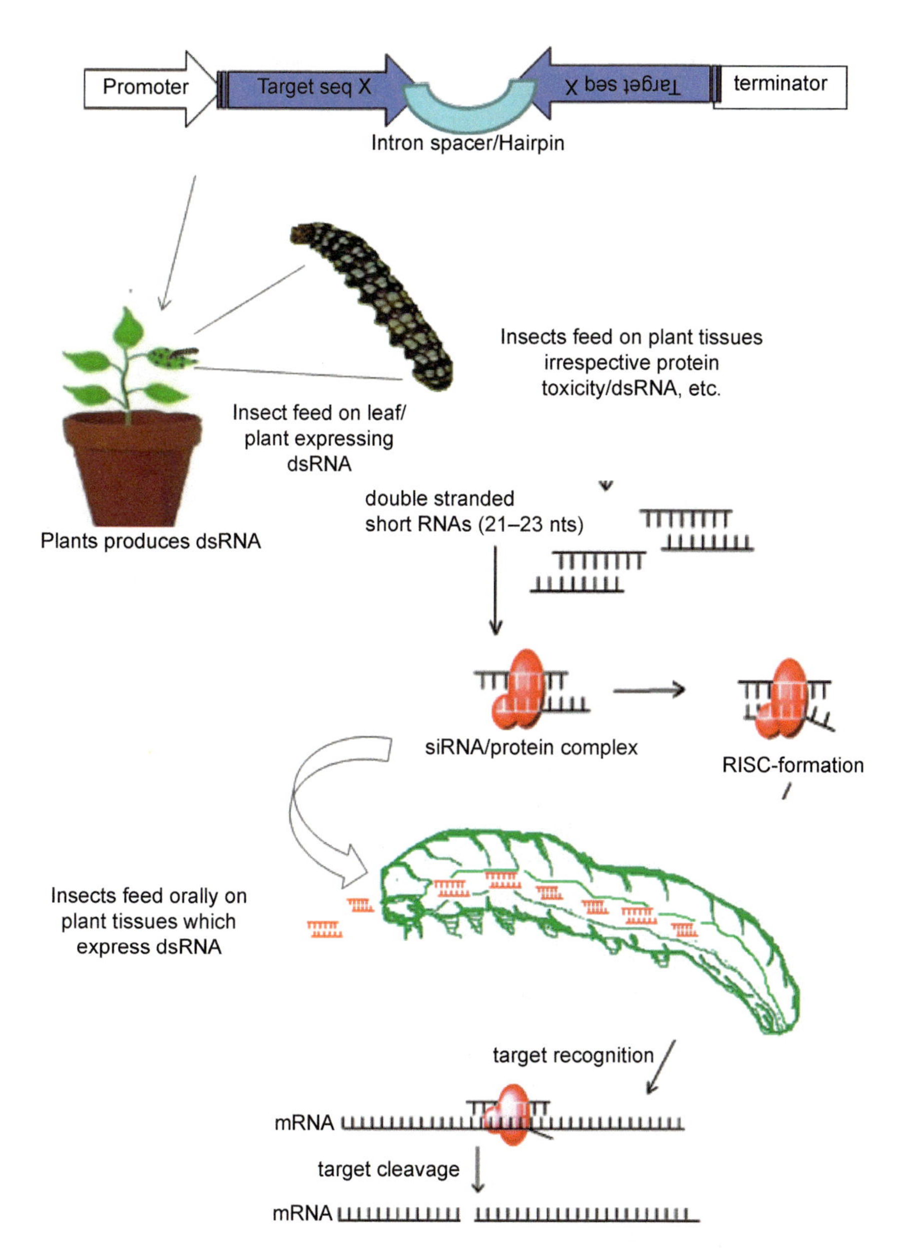

Figure 2. Mechanism of RNAi in insects. The dsRNA produced from the plants enters the insect midgut. The midgut epithelial cells recognize the dsRNA and cleaves the target mRNA and degrades it which results in reduced growth of larvae or death.

scale field studies to control insect pests. For efficient insect control, the organism should be able to autonomously take up the dsRNA by feeding onto dsRNA expressing leaf or other material. The insect gut determines the process of toxicity, uptake of nutrients and survival. Gut is divided into three regions; foregut, midgut and hindgut. The midgut is responsible for nutrient absorption, whereas excretion and water balance take place primarily in the Malpighian tubules attached to the hind end (Price and Gatehouse 2008). In the midgut nutrients absorb from the gut lumen with its large absorption area created by the microvilli, with many channels and endocytosis apparatus (Hakim et al. 2010).

Uptake of dsRNA in Insects

The efficiency of gene silencing in insects is strongly influenced by the uptake of dsRNA which also determines the potential for RNAi as an insect pest control agent. The major challenge for RNAi in insects is the delivery system because gene silencing is limited only to cells that uptake siRNAs (Terenius et al. 2011). The RNAi uptake mechanisms are divided into cell-autonomous and noncell-autonomous (Siomi and Siomi 2009). In case of cell-autonomous RNAi, the silencing process is limited to the cell in which the dsRNA is introduced/expressed and encompasses the RNAi process within the individual cell. In case of non-cell-autonomous RNAi, the gene silencing effect takes place in tissues/cells different from the location of application or production of the dsRNA. There are two different kinds of non-cell-autonomous RNAi: environmental RNAi and systemic RNAi (Whangbo and Hunter 2008). To control insects through gene silencing, non-cell-autonomous RNAi would better serve because the insect will internalize the dsRNA of a target gene through feeding. If the target gene is expressed in a tissue outside of the gut, the silencing signal will also have to spread via cells and tissues, which is systemic RNAi. Non-autonomous and systemic RNAi has been reviewed earlier in more detail (Voinnet 2005; Xie and Guo 2006; May and Plasterk 2005). In insects, physiology of RNAi uptake focused mainly on transmembrane channel and endocytosis mediated uptake mechanisms.

The Transmembrane Channel-mediated Uptake Mechanism

The best studied dsRNA uptake mechanism is in *Caenorhabditis elegans*. Fire et al. (1998) noticed that in *C. elegans*, the RNAi signal can spread from the injected cells to the surrounding tissue. The gene responsible for the uptake of dsRNA was identified in *C. elegans* named as *Systemic RNAi Defective* (*SID*). *SID1* is a transmembrane protein or channel that consists of eleven transmembrane domains and is mainly localized in cells that are in direct contact with the environment (Winston et al. 2002; Feinberg and Hunter 2003; Shih and Hunter 2011). It functions as a multimer transporting dsRNA passively into the *C. elegans* cells. However, it is not essential for the export of dsRNA from the cell (Winston et al. 2002). The other protein, SID-2, is mainly found in the intestine tissue of the worm and facilitates environmental RNAi (Winston et al. 2007). Three hypotheses are proposed regarding the relation/cooperation/coordination between these two proteins: (i) SID-2 modifies the SID-1 molecule to activate the transport, or (ii) it binds the dsRNA from the environment and delivers it to SID-1, or (iii) it induces the endocytosis pathway of the dsRNA, in which case SID-1 delivers the dsRNA to the cytoplasm (Whangbo and Hunter 2008). The possibility of other proteins involvement cannot be ruled out and there is scope to identify such transmembrane channel proteins in insects.

The Endocytosis-mediated Uptake Mechanism

In general, endocytosis internalizes molecules from outside the cell, which happens through vesicles budding inward from the plasma membrane. To internalize certain molecules, different ways exists based on the size of the molecules (Lodish et al. 2008). Phagocytosis, a form of endocytosis, plays a major role in the immune system of organisms and in the engulfment of apoptotic cells. Here the exogenous molecules are broken down by endo-proteases or peptidases. Cells can also engulf small molecules or liquids by another mechanism called as pinocytosis (Lodish et al. 2008). The pinocytosis consists of five steps; initiation, cargo selection, coat assembly, scission and un-coating (McMahon and Boucrot 2011). The cell can also internalize molecules or particles by receptor mediated endocytosis. When molecules bind to specific clathrin receptors on the cell surface, these regions bud inwards into a vesicle (Lodish et al. 2008). This pathway is not only used by the cell to internalize nutrients, but also to regulate the presence of proteins and receptors at the surface of the cell (McMahon and Boucrot 2011).

D. melanogaster has a cell-autonomous RNAi mechanism and hemocytes are able to respond to environmental RNAi (Roignant et al. 2003; Gordon and Waterhouse 2007; Miller et al. 2008). Screening of S2 cells of *D. melanogaster* soaked in dsRNA medium showed RNAi effect and alternate dsRNA uptake mechanisms (Saleh et al. 2006; Ulvila et al. 2006). Ulvila et al. (2006) tested 2,000 dsRNA fragments from a S2 cDNA library to identify RNAi pathway genes such as *Dicer 2* (*DCR2*) and *Argonaute 2-*(*Ago2*) from RNAi-induced lethality screening. This screening led to the identification of *Clathrin heavy chain* gene involved in cell-autonomous RNAi, a known component of the endocytosis machinery. The role of *vacuolar H+ ATPase* and combination of scavenger receptors participating in the uptake of dsRNA was confirmed by Saleh et al. (2006). Scavenger receptors are known to play an important role in the phagocytosis of bacterial pathogens (Erturk-Hasdemir and Silverman 2005; Kocks et al. 2005). While screening another set of S2 cDNA library, a total of 23 genes out of 7,216 were found to be involved in dsRNA uptake and processing (Saleh et al. 2006). Some of these sequences encode for proteins of the vesicle mediated transport, conserved oligomeric Golgi complex family, cytoskeleton organization and protein transport which are directly and/ or indirectly involved in endocytosis. These alternate uptake mechanisms in *C. elegans* provided greater insights in understanding the RNAi uptake mechanisms. Several studies suggest that the process of dsRNA uptake mechanism involved several different genes encoding receptors, adapters and sorting signal proteins. Targeting insect transport or receptor proteins for RNAi is also one of the strategies to control insects. In future, research efforts may provide evidences of using these targets for RNAi to combat insect pests.

Different Methods to Deliver dsRNA

The dsRNA delivery methods include microinjection (Bettencourt et al. 2002; Tomoyasu and Denell 2004; Ghanim et al. 2007), soaking, and oral feeding through artificial diet (Eaton et al. 2002; Turner et al. 2006; Baum et al. 2007; Mao et al. 2007; Chen et al. 2008; Tian et al. 2009). Microinjection has been effectively used to deliver short dsRNA to effectively inhibit the target gene function (Siomi and Siomi 2009). However, the steps involved are complicated and the injection pressure and wound created inevitably damage the insect skin and stimulate immune responses. The other limitation is the high cost for *in vitro* synthesis and storage of dsRNA. For few insects like *D. melanogaster*, the embryos are soaked in higher concentrations of dsRNA solution to inhibit gene expression (Eaton et al. 2002). Soaking *D. melanogaster* S2 cells in solutions of dsRNA derived from cell cycle genes such as *Cyclin E (CycE)* and *Ago* has been shown to effectively inhibit the expression of target genes (March and Bentley 2007). The soaking method of RNAi is rarely used in some insects because it is effective only in certain insect cells and tissues at particular developmental stages that absorb dsRNA from the solution.

Feeding of dsRNA through diet is the most attractive method of delivery because of its convenience and ease of manipulation. This mode of delivery is a natural method of introducing dsRNA into insect body and causes less damage to the insect when compared to microinjection (Chen et al. 2010). Feeding through diet is especially very useful in very small insects which are difficult to handle for micro injection. Even though, the early insect RNAi feeding studies were not a great success, later studies proved that this approach can be effectively used to silence genes. For example, feeding with dsRNA to silence aminopeptidase gene *Slapn* (*Spodoptera littoralis* aminopeptidase) did not achieve silencing in midgut of *Spodoptera littoralis*, but the injection of dsRNA effectively silenced this gene (Rajagopal et al. 2002). Feeding of dsRNA targeting *Epiphyas postvittana* carboxylesterase gene (*EposCXE1)* in the larval midgut and *E. postvittana* pheromone-binding protein *(EposPBP1)* in adult antennae successfully silenced the genes in larvae (Turner et al. 2006). Silencing of *Nitrophorin 2* (*NP2*) gene in salivary gland of *Rhodnius prolixus* through dsRNA feeding lead to shortened coagulation time of plasma (Araujo et al. 2006). Gene silencing by dsRNA feeding was also successful in several other insects, including insects of the orders Hemiptera, Coleoptera and Lepidoptera (Baum et al. 2007; Mao et al. 2007). Inspite of successful delivery of dsRNA through feeding, the major challenge was to ingest higher amounts of dsRNA (Chen et al. 2010). Upon ingestion of appropriate dsRNA, incomplete silencing of *Cellulase-1(CELL-1)* in termite *Reticulitermes flavipes* (Zhou et al. 2008), trehalose-6-phosphate synthase *(TPS)* in *N. lugens* nymphae (Chen et al. 2010), *NP2* in *R. prolixus* (Araujo et al. 2006) was observed. Another challenge is sensitivities of RNAi molecules when delivered orally vary with different insect species. The dsRNA could inhibit the expression of *Tsetse glutamic acid-proline repeat gene (TsetseEP)* in the midgut of *Glossina morsitans*, but could not inhibit the expression of the *Transferrin* gene *2A192* in fat bodies due to lack of transfer capacity between tissues (Walshe et al. 2009).

Several laboratory studies showed that delivery of dsRNA through different methods can be efficiently used to silence genes by RNAi in insects. Hence RNAi technology was extended for large scale control of insects by developing transgenic plants. Through transgenics the dsRNAs can be generated continuously and therefore stable dsRNA material can be obtained to effectively control insects. Even though yeast strains genetically engineered to synthesize dsRNA specific to *D. melanogaster* genes did not show successful gene silencing (Gura 2000), dsRNA produced in bacteria was effective against *C. elegans* (Timmons and Fire 1998). Lately, as described below, transgenic plants producing dsRNAs directed against gene function in Lepidoptera, Coleoptera, and Hemiptera pests have become more common (Gordon and Waterhouse 2007; Baum et al. 2007; Mao et al. 2007; Chen et al. 2010).

For effective RNAi to combat pest control the dsRNA delivery is crucial and RNAi has been shown to be a powerful tool in functional genomic research and demonstrated considerable potential. The dsRNA delivery and stability determines the efficiency of RNAi technology. Microinjection seems to be more accurate in delivering dsRNA to the target place of choice. Soaking insect cells is more convenient but transfection process may enhance uptake and delivery of dsRNA. Transfection reagents such as Cellfectin, Cellfectin II, Lipofectin, Lipofectamine 2000, ExGen 500, Metafectine TransFectin, DMRIE-C and Lipofectamine 2000 greatly enhanced the uptake and delivery of dsRNA. High transfection rates with CiE1 cells were reported by using Lipofectin as the transfection reagent (Johnson et al. 2010). Soaking *D. melanogaster* larvae with dsRNA solution showed 5%–8% reduction in target gene expression, however using four different transfection reagents (TransFectin, DMRIE-C, Cellfectin and Lipofectamine 2000) resulted in 31% to 52% reduction in target gene expression (Whyard et al. 2009).

Transgenic Approach to Control Insect Pests—Host Induced Gene Silencing

Plants evolved RNA silencing machineries to regulate developmental programs and to provide protection from invasion by foreign nucleic acids, such as viruses (Sidahmed and Wilkie 2010; Voinnet 2005). Silencing of several key pathogen mRNA by plant-expressed RNAi constructs showed resistance against important pathogens such as nematodes, fungi or aphids. This phenomenon is called as Host-Induced Gene Silencing (HIGS) and showed potential application in plant protection without the need of chemical disease control mechanisms. This natural phenomenon can be exploited to control agronomically relevant plant diseases, pest infestations. In plant pests and pathogens, such as insects (Price and Gatehouse 2008; Zhang et al. 2013), nematodes (Lilley et al. 2012) and fungi (Nunes and Dean 2012) the *in vitro* feeding of dsRNA can signal PTGS of target genes. The host-induced gene silencing (HIGS) is another potential promising alternative in plant protection because of its high selectivity for the target organism. However, the major bottleneck in successful HIGS strategy is the identification of suitable target genes in the infectious agent. Genes with known lethal knockout phenotypes can be highly efficient targets for HIGS. Effective targets also may be identified by screening cDNA libraries for highly expressed genes. It is critical to ensure that the dsRNA and corresponding siRNA species do not exert off-target effects that negatively impact host plant physiology, non-target host colonizers and/or mammals that feed on the modified crop. Several reports show successful development of transgenic plants expressing dsRNAs and effective silencing of insect genes resulted in inhibition of insect larvae growth or mortality. In this section, few successful examples where insect genes have been silenced by expressing insect specific dsRNA in plants are discussed. The systematic view of mechanism of dsRNA expression in plants and uptake of dsRNA and death of insect has been shown in Fig. 2.

In *C. elegans*, SID-1 and SID-2 intestinal transmembrane proteins have been shown to play critical roles in dsRNA uptake (Winston et al.

2002, 2007). The orthologs of SID-1 and SID-2 in many insects have been identified (but not all), suggesting conserved dsRNA uptake mechanisms in insects that could be exploited for HIGS (Gatehouse and Price 2011; Zhang et al. 2013). The first successful demonstration of HIGS is the dsRNA construct targeting the *Vacuolar H+ATPase* (*V-ATPase subunit* A) (Baum et al. 2007) against coleopteran Western corn rootworm (WCR; *Diabrotica virgifera*) in maize plants.

Baum et al. (2007) assessed WCR cDNA libraries to identify candidate target genes and dsRNA were prepared for testing in a WCR feeding assay. A total of 290 potential targets were identified and corresponding dsRNAs were synthesized *in vitro* and their effects on larval performance were determined by delivery through artificial diet. These assays identified 14 genes whose knockdown resulted in insect stunting and mortality 12 days after infestation. These genes encode Ubiquitin conjugating enzyme, Swelling-dependent chloride channel, Putative glyceraldehyde-3-phosphate dehydrogenase, Glucose-6-phosphate 1-dehydrogenase, Chitinase ortholog, Vacuolar-ATPase D subunit 1, ADP-ribosylation factor, Juvenile hormone esterase ortholog, Transcription factor IIB, Cytosolic juvenile hormone binding protein, V-ATPase A subunit 2, V-ATPase E, ATP synthase chain A, Mitochondrial ATPase6, Endoglucanase, ADP/ATP translocase, mRNA capping enzyme, Ribosomal protein L9, Ribosomal protein L19, 26S proteosome regulatory subunit p28, Putative β-tubulin, and Actin chromodomain helicase-DNA-binding protein. This study suggested that for efficient RNAi in insects, genes encoding essential functional proteins should be targeted. Amongst all these genes, transgenic corn expressing dsRNA targeting *V-ATPase* subunit was expressed in midgut. Transgenic corn plants showed protection against WCR infestation comparable to that of the *Bt* transgene (Baum et al. 2007). Other coleopteran species such as Southern corn rootworm (*Diabrotica undecimpunctata howardii*) and Colorado potato beetle (CPB; *Leptinotarsa decemlineata*) were tested for their sensitivity to dsRNAs in diet (Yang et al. 2013).

Insects have several large families of genes encoding enzymes (*P450s, Glutathione S-transferases and carboxylesterases*) which can detoxify plant chemical cocktails. Some of these have been implicated in the evolution of pesticide resistance, for example through tissue-specific transcriptional upregulation. Such genes are attractive targets for applying the ingestible dsRNA approach to manage resistance against plant allelochemicals or insecticides to make insect susceptible to those chemicals. Indeed, Mao et al. 2007 have corroborated the conclusion of Baum et al. 2007 that the strategy works for multiple midgut-expressed genes in rootworms by demonstrating silencing of a *Glutathione-S-transferase* gene in bollworm.

These findings strongly suggest that plant-delivered silencing of insect-detoxifying genes could be a powerful strategy for controlling insect pests.

The *Cytochrome P450* (*CYP6AE14*) gene was identified in cotton bollworm (*H. armigera*), which confers resistance against the cotton metabolite gossypol that is toxic to cotton bollworm. *CYP6AE14* is highly induced in the midgut and expression levels correlate with larval growth when gossypol is added into the insect diet. Transgenic cotton plants were developed by expressing *CYP6AE14* RNAi construct and *H. armigera* larvae were fed with plant material expressing double-stranded RNA (dsRNA) for *CYP6AE14*. The transcript levels of *CYP6AE14* were significantly reduced in the midgut. The larval growth was reduced and mortality was increased in the presence of gossypol (Mao et al. 2007). The *CYP6AE14* dsRNA expressing transgenic Arabidopsis, tobacco or cotton showed reduced expression of this gene and enhanced sensitivity to gossypol and larvae reared on these plants showed reduced growth and mortality (Mao et al. 2007, 2011). The co-expression of plant cysteine proteases with *CYP6AE14* dsRNA showed enhanced level of bollworm protection (Mao et al. 2013).

The plant cysteine proteases (CP), such as *GhCP1* from cotton and *AtCP2* from Arabidopsis attenuate the cotton bollworm midgut peritrophic matrix, which presumably improves transmission of dsRNA from the plant to midgut cells. HIGS targeting the other *cytochrome P450* genes were used to combat cotton bollworm resistance to pyrethroid insecticides. The dsRNA targeting *CYP9A14* is Arabidopsis transgenics showed low larval resistance to deltamethrin, a widely used pyrethroid insecticide for cotton (Tao et al. 2012). Injection of dsRNAs against *CYP6B7, cytochrome P450 reductase or cytochrome b5* into the midgut of cotton bollworm larvae led to restoration of susceptibility to the pyrethroid insecticide fenvalerate (Tang et al. 2012).

Using microarray in soybean cyst nematode (SCN; *Heterodera glycines*) and *C. elegans*, 32 genes that showed high homology to *C. elegans* genes having lethal mutant or RNAi phenotypes also showed high expression in SCN parasitic stages and candidate genes that cause insect lethality when silenced were identified. The defence pathway genes were induced even at early (by 6 hours' post infection) stages of compatible reaction. The defence related genes such as *Kunitz trypsin inhibitor (KTI), germin, peroxidase, phospholipase D, 12-oxyphytodienoate reductase (OPR), pathogenesis related-1 (PR1), phospholipase C, lipoxygenase, WRKY6 transcription factor* and *calmodulin* showed 80–90% reduction in SCN females reaching maturity (Alkharouf et al. 2007; Klink et al. 2009). The HIGS targeting *H. glycines* aldolase (*HgALD*), which encodes fructose-1,6-diphosphate aldolase, a key enzyme of gluconeogenesis showed 58% reduction in mature females (Youssef et al. 2013). HIGS is also effective against parasitism in root knot nematodes (RKN). The *Arabidopsis* transgenic plants expressing RNAi construct targeting conserved RKN effector gene *16D10* encodes a small secretory peptide that helps establish feeding sites, conferred broad-spectrum resistance to the four major RKN species (Huang et al. 2006). Similarly, *16D10* targeting transgenic grape hairy roots expressing hairpin-based silencing constructs showed reduced susceptibility to *Meloidogyne incognita* (Yang et al. 2013).

A dsRNA-based *in vivo* delivery system was tested on the model organism *R. flavipes*, the most common subterranean termite in North America. In an effort to test target genes with pest control potential, a *nutrition-related cellulase* gene (*Cell-1*) and two *soldier-inhibitory hexamerin* genes (*Hexa-1 & 2*) were checked. Suppression of these target genes impacted vital biological processes of termites including but not limited to feeding, juvenile hormone modulation, development, and differentiation (Bakhetia et al. 2005). Eventually, these biological impacts led to individual death and/or compromised fitness of the insects.

Xiong et al. (2013) developed transgenic tobacco (*Nicotiana tabacum*) overexpressing hairpin RNA (hpRNA) targeting *H. armigera hormone receptor 3* (*HaHR*3). HR3 is a molt-regulating transcription factor which plays a key role during metamorphosis by regulating metamorphosis-related gene expression. The transgenic tobacco plants expressing dsRNA against *HaHR3* gene showed significantly reduced transcript levels of *HaHR3* and upon feeding of leaves from these transgenic plants resulted in developmental deformity and larval death of *H. armigera*.

The invertebrate arginine kinase (AK) is one of the potential targets to combat insects. The feeding bioassays on the dsRNA of *Phyllotreta striolata AK* produced by *in vitro* transcription at minute quantities caused impaired development in beetle (Zhao et al. 2008). AK could be a potential target because it is not present in vertebrates and the biosynthetic pathway of phosphoarginine is completely different from mammalian tissues (Pereira et al. 2005; Wu et al. 2007). In a recent study, Liu et al. (2015) silenced the *AK* gene of *H. armigera* (*HaAK*). AK is a phosphotransferase involved in reversible catalysis of phosphate from ATP to arginine and yielding phosphoarginine and ADP. This process plays a critical role in cellular energy metabolism in invertebrates. Leaves of transgenic *Arabidopsis* plants expressing *HaAK* dsRNA, when fed by *H. armigera,* showed significantly retarded larval growth and 55% mortality.

RNAi has been effectively used to understand gene function in some insect species. Hemolin from the giant silkmoth (*Hyalophora cecropia*) identified as a bacteria-inducible molecule and a member of the immunoglobulin super family, is present in oocytes and embryos. Bettencourt et al. 2002 used RNAi to investigate *Hemolin* gene function in *H. cecropia* and demonstrated its role in normal development of embryos.

When RNAi-females were mated, no larvae emerged from their eggs and when dissected the eggs revealed malformed embryos. In *Bombyx mori* (silkworm; Quan et al. 2002), injection of dsRNA, corresponding to the *White egg* gene (*BmWh3*), into preblastoderm eggs of the wild-type silkworm induced phenotypes similar to those observed with *wh3* mutants characterized by white eggs and translucent larval skin. Similarly, in grasshopper injection of dsRNA corresponding to the eye color gene *Vermilion* into first instar nymphs triggered suppression of ommochrome formation in the eye lasting through two instars, equivalent to 10–14 days in absolute time. These results suggest that systemic dsRNA application elicits specific and relatively long term gene silencing in juvenile grasshopper instars (Dong and Friedrich 2005).

Fabrick et al. (2004), demonstrated that RNAi can also be utilized in the lepidopteran, *Plodia interpunctella*. A cDNA for *Tryptophan oxygenase* was cloned and it was seen that silencing of this gene through RNAi during embryonic development resulted in loss of eye-color pigmentation in lepidopteran insects. In addition to all the genes mention in this section, several other genes have been successfully targeted by RNAi in insects as shown in Table 1.

MicroRNA Mediated Insect Control Strategies

MicroRNAs (miRs) are non-coding small RNAs (18–25 nt) that play a crucial role in various biological processes including development as regulators of gene expression. The functional mechanisms of miRs in regulating all intrinsic cellular functions have been studied in several organisms. MiRs function through post transcriptional regulation by binding to complementary regions of messenger RNA (mRNA) in the 3′ untranslated region (UTR), 5′ UTR and the coding region and play critical roles in many physiological processes like growth, development, metabolism, behaviour and apoptosis (Hussain and Asgari 2010; Bartel 2007; Giraldez et al. 2006; Yu et al. 2007). By using computational or experimental methods, more than 30,000 miRs from different living organisms such as *Spodoptera litura*, *Bombyx mori*, *C. elegans*, *Arabidopsis thaliana* and *Homo sapiens* have been identified. Sequences of all these miRs are available in the miRBase site (http://www.mirbase.org/).

In insects, a total of about 2300 miRs have been identified from 22 insect species including *D. melanogster*, *Anopheles gambie*, *Apis melifera*, *B. mori* and *D. pseudoobscura* and deposited in miRBase. However, large sets of insect miRs were identified by computational method with a few reports showing the importance of miRs in developmental stages, metabolism and in response to viral infection from polyphagous *Spodoptera* sp. (Rao et al. 2012; Mehrabadi et al. 2013).

Table 1. Insect genes targeted by RNAi.

Gene	Insect	Protein encoded/function	Mode of dsRNA delivery	Effect on insect	Reference
β-Actin	*Spodoptera littoralis*	Beta Actin—Component of cytoskeleton	Injection	Sperm release was disrupted	Gvakharia et al. 2003
TcCHS1	*Tribolium castaneum*	Chitin synthase—chitin deposition	Injection	Moulting of larva, larval-pupal and pupal-adult was disrupted. Reduction in whole body chitin content	Arakane et al. 2005
TcCHS2				Cessation of feeding, reduced larval size and decreased chitin content in the midgut	
SeCHS gene	*Spodoptera exigua*	Chitin synthase—chitin deposition in the eggshell and development of pharynx	Injection	Disorder in the insect cuticle, no expansion of the larval trachea epithelial wall and other larval abnormalities	Chen et al. 2008
Eye colour gene vermilion	*Schistocerca americana*	Tryptophan oxygenase-enzyme participates in tryptophan metabolism	Injection	Suppression of ommochrome formation and systematic expression	Dong and Friedrich 2005
Ar white	*Athalia rosae*	ATP-binding cassette (ABC) transporter	Injection	White phenocopy in embryonic eye pigmentation	Sumitani et al. 2005
BgRXR gene	*Blattella germanica*	Retinoid X receptor—Nuclear receptor which binds 20-hydroxyecdysone	Injection	Inhibition of pupal eclosion	Martin et al. 2006

Table 1 contd. ...

...Table 1 contd.

Gene	Insect	Protein encoded/function	Mode of dsRNA delivery	Effect on insect	Reference
Relish	*Apis mellifera*	NF-kappaB- and IkappaB-related proteins—Transcription factor activity	Injection	Inhibition of *Relish* gene expression and reduction in the expression of two other immune genes, *Abaecin* and *Hymenoptaecin*	Schlüns and Crozier 2007
Per	*Gryllus bimaculatus*	Period circadian protein homolog 1—Circadian clock function	Injection	Complete loss of circadian control of locomotor activity and electrical activity in the optic lobe	Moriyama et al. 2008
Per	*Spodoptera littoralis*	Period circadian protein homolog 1—Circadian clock function	Injection	Delayed sperm release	Kotwica et al. 2009
MaLac2	*Monochamus alternatus*	Laccase-catalyze ring cleavage of aromatic compounds	Injection	Pupal and adult cuticle sclerotisation, death at a high dose	Niu et al. 2008
TcCHT5	*Tribolium castaneum*	Chitinase-like proteins—chitin deposition	Injection	Pupal-adult moulting and adult eclosion was affected	Zhu et al. 2008
TcCHT10				Egg hatching, larval moulting, pupation and adult metamorphosis were affected	
TcCHT7				Abnormalities in abdominal contraction and wing/elytra extension	
Pdf	*Blattella germanica*	Pigment dispersing factor—dispersing factor gene	Injection	Effects on insect night activity	Lee et al. 2009

NOS	*Gryllus bimaculatus*	Nitric oxide synthase—catalyzing the production of nitric oxide (NO) from L-arginine		Destruction of long-term memory	Takahashi et al. 2009
HaHMGR	*H. armigera*	3-hydroxy-3-methylglutaryl coenzyme A reductase—key enzyme in mevalonate pathway	Injection	Inhibits oviposition	Wang et al. 2013
EposCXE1 and EposPBP1	*Epiphyas postvittana*	Carboxylesterase and pheromone binding protein—involved in the initiation, movement and amplification of the silencing signal	Oral Feeding	Inhibition of gene expression, reduces adult antennae in insects	Turner et al. 2006
CYP6AE14	*Helicoverpa armigera*	Cytochrome P450 6AE14—catabolism of gossypol	Oral Feeding	Inhibition of larval growth	Mao et al. 2007
GST1		Glutathione-S-transferase—catalyze the conjugation of the reduced form of glutathione		Successful inhibition of gene expression	
v-ATP	*Diabrotica virgifera virgifera* LeConte	Vacuolar- ATPase—acidify a wide array of intracellular organelles and pump protons across the plasma membranes	Oral Feeding	Delayed larval development and increased mortality	Baum et al. 2007
AK	*Phyllotreta striolata*	Arginine kinase—Maintenance of ATP levels by the phosphorylation	Oral Feeding	Retarded development, increased mortality and reduced fertility	Zhao et al. 2008

Table 1 contd. ...

...Table 1 contd.

Gene	Insect	Protein encoded/function	Mode of dsRNA delivery	Effect on insect	Reference
Cell-1	*Reticulitermes flavipes*	Cellulase enzyme—hydrolyze cellulose	Oral Feeding	Reduction in group fitness and increased mortality	Zhou et al. 2008
Hex-2		Hexamerin—Caste regulatory hexamerin storage protein			
TsetseEP gene	*Glossina morsitans*	Immuno responsive midgut expressed gene	Oral Feeding	Inhibition of *TsetseEP* gene expression, but no inhibition of *2A192* gene expression—causes insect mortality	Walshe et al. 2009
Transferrin gene 2A192		Iron-binding blood plasma glycoproteins			
NlTPS	*Nilaparvata lugens*	Trehalose phosphate Synthase—Trehalose Biosynthesis	Oral Feeding	Disturbed development through disruption in the TPS enzymatic activity, reduction of insect survival rate	Chen et al. 2010
V-ATPase A	western corn rootworm	Vacuolar ATPase—acidify a wide array of intracellular organelles and pump protons across the plasma membranes	Transgenic corn	A significant reduction in western corn rootworm feeding damage	Baum et al. 2007
CYP6AE14	*H. armigera*	Cytochrome P450 gene—catabolism of gossypol	Transgenic *Arabidopsis thaliana* and tobacco	Significant retardation of larval growth	Mao et al. 2007
CYP6AE14	*H. armigera*	Cytochrome P450—catabolism of gossypol	Transgenic Cotton	Significant retardation of larval growth	Mao et al. 2011

NlHT1	*Nilaparvata lugens*	Hexose transporter–transporter	Transgenic Rice	Reduced transcripts of the targeted genes in midgut	Zha et al. 2011
Nlcar		Carboxypeptidase—involved in the digestion of food			
Nltry		Trypsin-like serine protease—cleave peptide bonds following a positively charged amino acid (lysine or arginine)			
Rack	*Myzus persicae*	Receptor of activated protein kinase C—involved in small interfering RNA induced apoptosis associated with activation of caspases	Transgenic *N. benthamiana, A. thaliana*	Silenced aphids have reduced progeny production	Pitino et al. 2011
HaEcR	*H. armigera*	Ecdysone receptor-signalling	Transgenic tobacco	Significant molting defects and lethality	Zhu et al. 2012
HaHR3	*H. armigera*	Orphan nuclear receptor HR3—Molt-regulating transcription factor	Transgenic Tobacco	Developmental deformity of larvae and lethality	Xiong et al. 2013
HaAK	*H. armigera*	Arginine kinase—Maintenance of ATP levels by the phosphorylation	Transgenic *Arabidopsis*	Retardation of larval growth and mortality	Liu et al. 2015

In closely related species the miRs are found to be highly conserved in different tissues and developmental stages. However, the uniqueness of the organism to evolutionary divergence depends on genus specific miRs. In insects, the evolutionary sequence conservation has become a hallmark of miR biology with the identification of new miRs (Ruby et al. 2007; Lai et al. 2003). Species specific miRs exist, reflecting system-specific characteristics and certain miRs are conserved across insect, mammals and distinctly apparent phylogenetic relationship found within insects.

Recently in *Spodoptera frugiperda* cell line Sf21, a total of 226 miRNAs were identified. From these identified miRNAs, 116 are highly conserved in other insects, like *B. mori*, *D. melanogaster* and *Tribolium castenum*. Based on comparative analysis with the insect miRNA data sets 110 miRNAs are considered as novel and five miRNA clusters were identified and the largest one encodes five miRNA genes. The expression of 12 miRs; seven known and five novel miRs were analysed in Sf21 cell lines. These miRs (*sfr-mir-305-5p, sfr-mir-307-3p, sfr-mir-71-3p, sfr-mir-281, sfr-mir-317, sfr-mir-2756, sfr-mir-932, sfr-mir-184-3p, sfr-mir-2766, Novel_miR15, Novel_miR16, and Novel_miR17*) showed expression in Sf21 cells. Amongst the identified miRs, *sfr-mir-305-5p* was highly expressed as compared to the other known and novel miRs. Based on bioinformatic analysis, miRs from *S. frugiperda* are highly homologous to *B. mori* miRs compared to other insects such as *D. melanogaster*, *T. castenum*, *A. aegypti* and *A. gambie* (Kakumani et al. 2015).

In insects, the first report showing utilization of synthetic inhibitor that mimic miR *har-miR-2002b* was against trypsin like serine protease to inhibit the growth of *H. armigera*. The bioassays by oral feeding indicated 75% reduction in fecundity and 40% larval mortality in the presence of har-miR-2002b (Jayachandran et al. 2013). With the progress of identifying new miRs, identifying the specific targets is crucial to utilize this technology in agriculture to combat insect pests. This technology has the potential to be used as alternate pest management tactics.

Future Prospects of RNAi in Insects

Despite the success and effectiveness of RNAi in insects, researchers are consciously aware of unpredictable issues related with RNAi in insects. The efficacy may vary with target genes, mode of action, delivery, across the taxa and between different laboratories and researchers (Terenius et al. 2011). However, at present, limited resources and ability to predict the ideal experimental strategy for RNAi for a particular target gene and insect are not available, because of sparse understanding of RNAi signal amplification and spread among the insect cells. The systematic analysis at molecular physiology of insects and RNAi mechanisms is highly essential to overcome these limitations.

The fundamental biological question is how to improve the efficacy of RNAi application between different organisms and different life stages of insects. The success of RNAi depends on the genetic backgrounds (Kitzmann et al. 2013), cellular uptake and propagation of signal (Roignant et al. 2003; Miller et al. 2008) and presence or absence of the core RNAi machinery (Arimatsu et al. 2007). In mosquitoes it has been emphasized that the effect of RNAi on insect viruses is dependent on the temperature, RNA silencing in disease vector mosquitoes is efficient at low temperature (18°C). By comparing both transgenic and non-transgenic mosquitoes the direct or indirect inhibition of AGO-2-dependent silencing may be the reason for low efficacy silencing (Adelman et al. 2013). It is noteworthy to mention that temperature dependent defects in RNAi have also been described in many plants as well (Szittya et al. 2003; Chellappan et al. 2005).

The differences in tissue specific RNAi efficacy can be overcome by designing new efficient delivery methods such as viral transduction or transgenic development under constitutive promoters or tissue specific promoters depending on the gene which is targeted.

The success of RNAi technology depends on identification and preparation of specific RNAi molecule (in the form of dsRNA, siRNA, or a hairpin RNA) for a target gene of interest (GOI) of the insect. The progress in high throughput sequencing facilitates the target gene selection, which should be important to the target insects but friendly to nontarget animals and humans (Wang et al. 2011).

The length of dsRNA also plays a critical role in efficient RNAi. The length of the dsRNA used determines the efficiency of RNAi, usually 300–520 bp were used with a variation from 134–1824 bp. The minimal length required to obtain maximal biological activity varies among insect species, but generally > 50–200 bp length dsRNA yielded greater success with insect RNAi (Huvenne and Smagghe 2010; Bolognesi et al. 2012). In *Acyrthosiphon pisum* the RNAi trigger designed against the 5′ or 3′ end of the gene against hunchback (hb) resulted in insect mortality (Mao and Zeng 2012), but in contrast, in *Aedes aegypti*, dsRNA targeting the inhibitor of apoptosis gene, the 3′ end yielded a greater effect on mosquito mortality than 5′ or central region of the gene (Pridgeon et al. 2008).

It is highly imperative to screen multiple RNAi sequences for a gene of interest. siRNAs synthesized directly or obtained by dicing the dsRNA *in vitro* before administration to the insect also showed gene silencing in lepidopteran *H. armigera* (Kumar et al. 2012), aphid *A. pisum* (Mutti et al. 2006) and tsetse fly (Attardo et al. 2012). This suggests that sometime the shorter RNAi molecule may be ideal to obtain specificity. This phenomenon may be highly useful where the members of a gene family have high sequence similarity. Two genes with high sequence similarities can be silenced by the same dsRNA (Zhang et al. 2010).

In field applications, it is extremely important to design RNAi molecule for a exquisite target gene of interest which should be highly species specific. It is evident in *Drosophila* feeding species-specific *vATPase* dsRNA resulted in enhanced mortality in four conspecific species (species specific), but not in heterospecific flies (Whyard et al. 2009). This property may allow the researchers to use the allelic variants to silence the desired gene. In contrast, in genetically-distinct heterozygous individuals two alleles within a insect population can be targeted by RNAi.

Greater conceptual understanding of RNAi function in insects will facilitate the application of RNAi for dissection of gene function as well as fast-track application of RNAi to control pests. The transgenic approach expressing dsRNA against insect/host genes is shown to be effective. This technology can be effectively employed with careful analysis of off target genes and precise designing of dsRNA to a target gene against species of interest.

Acknowledgements

KSM laboratory projects are supported by the Samuel Roberts Noble Foundation, National Science Foundation and Gates Foundation. Travel and stipend to RSV and KCB were supported by Fulbright-Nehru Postdoctoral Fellowships by USIEF, India.

References

Adelman, Z.N., A.E.A. Michelle, R.W. Michael et al. 2013. Cooler temperatures destabilize RNA interference and increase susceptibility of disease vector mosquitoes to viral infection. PLoS Negl Trop Dis. 7(5): e2239.

Alkharouf, N.W., V.P. Klink and B.F. Matthews. 2007. Identification of *Heterodera glycines (soybean cyst nematode* [*SCN*]) cDNA sequences with high identity to those of *Caenorhabditis elegans* having lethal mutant or RNAi phenotypes. Exp Parasitol. 115: 247–258.

Arakane, Y., S. Muthukrishnan, K.J. Kramer et al. 2005. The *Tribolium* chitin synthase genes *TcCHS1* and *TcCHS2* are specialized for synthesis of epidermal cuticle and midgut peritrophic matrix. Insect Mol Biol. 14: 453–463.

Araujo, R.N., A. Santos, F.S. Pinto et al. 2006. RNA interference of the salivary gland nitrophorin 2 in the triatomine bug *Rhodnius prolixus (Hemiptera: Reduviidae)* by dsRNA ingestion or injection. Insect Biochem Mol Biol. 36: 683–693.

Arimatsu, Y., E. Kotani, Y. Sugimura et al. 2007. Molecular characterization of a cDNA encoding extracellular dsRNase and its expression in the silkworm, *Bombyx mori*. Insect Biochem Mol Biol. 37: 176–183.

Asgari, S. 2013. MicroRNA functions in insect. Insect Biochem Mol Biol. 43: 388–397.

Attardo, G.M., J.B. Benoit, V. Michalkova et al. 2012. Analysis of lipolysis underlying lactation in the tsetse fly, *Glossina morsitans*. Insect Biochem Mol Biol. 42: 360–370.

Bakhetia, M. 2005. RNA interference and plant parasitic nematodes. Trends Plant Sci. 10: 362–367.

Bartel, D.P. 2007. MicroRNAs: Genomics, biogenesis, mechanism, and function. Cell. 131: 11–29.

Bartel, D.P. 2009. MicroRNAs: target recognition and regulatory functions. Cell. 136: 215–233.

Baum, J.A., T. Bogaert, W. Clinton et al. 2007. Control of coleopteran insect pests through RNA interference. Nat Biotechnol. 25: 1322–1326.

Belles, X. 2010. Beyond *Drosophila*: RNAi *in vivo* and functional genomics in insects. Ann Rev Entomol. 55: 111–128.

Benbrook, C.M. 2012. Impacts of genetically engineered crops on pesticide use in the U.S.—the first sixteen years. Env Sci Europe. 24: 24.

Bettencourt, R., O. Terenius and I. Faye. 2002. *Hemolin* gene silencing by ds-RNA injected into Cecropia pupae is lethal to next generation embryos. Insect Mol Biol. 11: 267–271.

Bolognesi, R., P. Ramaseshadri, J. Anderson et al. 2012. Characterizing the mechanism of action of double-stranded RNA activity against western corn rootworm (*Diabrotica virgifera virgifera* LeConte). PLoS One. 7: e47534.

Chellappan, P., R. Vanitharani, F. Ogbe et al. 2005. Effect of temperature on geminivirus-induced RNA silencing in plants. Plant Physiol. 138: 1828–1841.

Chen, J., D. Zhang, Q. Yao et al. 2010. Feeding-based RNA interference of a trehalose phosphate synthase gene in the brown planthopper, *Nilaparvata lugens*. Insect Mol Biol. 19: 777–786.

Chen, X., H. Tian, L. Zou et al. 2008. Disruption of *Spodoptera exigua* larval development by silencing chitin synthase gene A with RNA interference. Bulletin Entomol Res. 98: 613–619.

Cogoni, C., J.T. Irelan, M. Schumacher et al. 1996. Transgene silencing of the al-1 gene in vegetative cells of Neurospora is mediated by a cytoplasmic effector and does not depend on DNA-DNA interactions or DNA methylation. EMBO J. 15: 3153–3163.

Dong, Y. and M. Friedrich. 2005. Nymphal RNAi: systemic RNAi mediated gene knockdown in juvenile grasshopper. BMC Biotechnol. 5: 25.

Eaton, B.A., R.D. Fetter and G.W. Davis. 2002. Dynactin is necessary for synapse stabilization. Neuron. 34: 729–741.

Erturk-Hasdemir, D. and N. Silverman. 2005. Eater: a big bite into phagocytosis. Cell. 123: 190–192.

Fabrick, J.A., M.R. Kanost and J.E. Baker. 2004. RNAi-induced silencing of embryonic tryptophan oxygenase in the pyralid moth, *Plodia interpunctella*. J Insect Sci. 4: 15.

Feinberg, E.H. and C.P. Hunter. 2003. Transport of dsRNA into cells by the transmembrane protein SID-1. Science. 301: 1545–1547.

Ferré, J. and J. Van Rie. 2002. Biochemistry and genetics of insect resistance to *Bacillus thuringiensis*. Annu Rev Entomol. 47: 501–533.

Fire, A., S. Xu, M.K. Montgomery et al. 1998. Potent and specific genetic interference by double-stranded RNA in *Caenorhabditis elegans*. Nature. 391: 806–811.

Fire, A.Z. 2007. Gene silencing by double-stranded RNA (Nobel lecture). Angewandte Chemie-International Edition. 46: 6967–6984.

Gatehouse, J.A. 2011. Prospects for using proteinase inhibitors to protect transgenic plants against attack by herbivorous insects. Curr Protein Pept Sci. 12: 409–416.

Gatehouse, J.A. and D.R.G. Price. 2011. Protection of crops against insect pests using RNA interference. *In*: Vilcinskas, A. (ed.). Insect Biotechnology, Biologically-Inspired Systems. 2: 145–168.

Ghanim, M., S. Kontsedalov and H. Czosnek. 2007. Tissue-specific gene silencing by RNA interference in the whitefly *Bemisia tabaci* (Gennadius). Insect Biochem Mol Biol. 37: 732–738.

Ghildiyal, M. and P.D. Zamore. 2009. Small silencing RNAs: an expanding universe. Nat Rev Genet. 10(2): 94–108.

Giraldez, A.J., Y. Mishima, J. Rihel et al. 2006. *Zebrafish* MiR-430 promotes deadenylation and clearance of maternal mRNAs. Science. 312: 75–79.
Gordon, K.H.J. and P.M. Waterhouse. 2007. RNAi for insect-proof plants. Nat Biotech. 25: 1231–1232.
Gura, T. 2000. A silence that speaks volumes. Nature. 404(6780): 804–808.
Gvakharia, B.O., P. Bebas, B. Cymborowski et al. 2003. Disruption of sperm release from insect testes by *cytochalasin* and *beta-actin* mRNA mediated interference. Cell Mol Life Sci. 60(8): 1744–1751.
Hakim, R.S., K. Baldwin and G. Smagghe. 2010. Regulation of midgut growth, development, and metamorphosis. Ann Rev of Ent. 55: 112408–085450.
Huang, G., R. Allen, E.L. Davis et al. 2006. Engineering broad root-knot resistance in transgenic plants by RNAi silencing of a conserved and essential root-knot nematode parasitism gene. Proc Natl Acad Sci USA. 26: 14302–14306.
Hussain, M. and S. Asgari. 2010. Functional analysis of a cellular microRNA in insect host-ascovirus interaction. J Virol. 84: 612–620.
Huvenne, H. and G. Smagghe. 2010. Mechanisms of dsRNA uptake in insects and potential of RNAi for pest control: a review. J Ins Physiol. 56(3): 227–235.
Jayachandran, B., M. Hussain and S. Asgari. 2013. An insect trypsin-like serine protease as a target of microRNA: utilization of microRNA mimics and inhibitors by oral feeding. Ins Biochem Mol Biol. 43(4): 398–406.
Johnson, J.A., K. Bitra, S. Zhang et al. 2010. The *UGA-CiE1* cell line from *Chrysodeixis includens* exhibits characteristics of granulocytes and is permissive to infection by two viruses. Ins Biochem Mol Biol. 40: 394–404.
Kakumani, P.K., M. Chinnappan, K.S. Ashok et al. 2015. Identification and characteristics of microRNAs from *Army Worm, Spodoptera frugiperda* cell line Sf21. PLoS ONE. 10(2): e0116988.
Ketting, R.F. 2011. The many faces of RNAi. Dev Cell. 20: 148–161.
Kitzmann, P., J. Schwirz, C. Schmitt-Engel et al. 2013. RNAi phenotypes are influenced by the genetic background of the injected strain. BMC Genomics. 14:5.
Klink, V.P., K.H. Kim, V. Martins et al. 2009. A correlation between host-mediated expression of parasite genes as tandem inverted repeats and abrogation of development of female *Heterodera glycines* cyst formation during infection of Glycine max. Planta. 230: 53–71.
Kocks, C., J.H. Cho, N. Nehme et al. 2005. Eater, a transmembrane protein mediating phagocytosis of bacterial pathogens in Drosophila. Cell. 123: 335–346.
Kotwica, J., P. Bebas, B.O. Gvakharia et al. 2009. RNA interference of the period gene affects the rhythm of sperm release in moths. J Biol Rhythms. 24(1): 25–34.
Kumar, P., S.S. Pandit and I.T. Baldwin. 2012. Tobacco rattle virus vector: a rapid and transient means of silencing *Manduca sexta* genes by plant mediated RNA interference. PLoS One. 7: e31347.
Kumar, S., A. Chandra and K.C. Pandey. 2008. *Bacillus thuringiensis* (*Bt*) transgenic crop: an environment friendly insect-pest management strategy. J Environ Biol. 29(5): 641–53.
Lai, E.C., P. Tomancak, R.W. Williams et al. 2003. Computational identification of *Drosophila* micro-RNA genes. Genome Biol. 4: R42.
Lee, C.M., M.T. Su and H.J. Lee. 2009. Pigment dispersing factor: an output regulator of the circadian clock in the German cockroach. J Biol Rhythms. 24(1): 35–43.
Lilley, C.J., L.J. Davies and P.E. Urwin. 2012. RNA interference in plant parasitic nematodes: a summary of the current status. Parasitology. 139: 630–640.
Liu, F., X.D. Wang, Y.Y. Zhao et al. 2015. Silencing the *HaAK* Gene by transgenic plant-mediated RNAi impairs larval growth of *Helicoverpa armigera*. Int J Biol Sci. 11(1): 67–74.
Lodish, H., A. Berk, C.A. Kaiser et al. 2008. Molecular cell biology. Chapter 9: Visualizing, Fractionating and Culturing Cells. W.H. Freeman and Company, Sixth Edition.
Mao, J. and F. Zeng. 2012. Feeding-based RNA interference of a *gap* gene is lethal to the pea aphid, *Acyrthosiphon pisum*. PLoS One. 7: e48718.

Mao, Y.B., W.J. Cai, J.W. Wang et al. 2007. Silencing a cotton bollworm *P450 monooxygenase* gene by plant-mediated RNAi impairs larval tolerance of gossypol. Nat Biotech. 25(11): 1307–1313.

Mao, Y.B., X.Y. Tao, X.Y. Xue et al. 2011. Cotton plants expressing *CYP6AE14* double-stranded RNA show enhanced resistance to bollworms. Trans Res. 20: 665–673.

Mao, Y.B., X.Y. Xue, X.Y. Tao et al. 2013 Cysteine protease enhances plant-mediated bollworm RNA interference. Plant Mol Biol. 83: 119–129.

March, J.C. and W.E. Bentley. 2007. RNAi-based tuning of cell cycling in *Drosophila* S2 cells: Effects on recombinant protein yield. Appl Microbiol Biotechnol. 73(5): 1128–1135.

Martin, D., O. Maestro, J. ruz et al. 2006. RNAi studies reveal a conserved role for *RXR* in molting in the cockroach *Blattella germanica*. J Insect Physiol. 52(4): 410–416.

Matranga, C. and P. Zamore. 2007. Small silencing RNAs. Curr Biol. 17: R789–R793.

May, R.C. and R.H. Plasterk. 2005. RNA interference spreading in *C. elegans*. Methods Enzymol. 392: 308–315.

McMahon, H.T. and E. Boucrot. 2011. Molecular mechanism and physiological functions of clathrin-mediated endocytosis. Nat Rev Mol Cell Biol. 12(8): 517–533.

Mehrabadi, M., M. Hussain and S. Asgari. 2013. MicroRNAome of *Spodoptera frugiperda* cells (Sf9) and its alteration following *Baculovirus* infection. J Gen Virol. 94: 1385–1397.

Miller, S.C., S.J. Brown and Y. Tomoyasu. 2008. Larval RNAi in *Drosophila*? Development Genes and Evolution. 218: 505–510.

Moriyama, Y., T. Sakamoto, S.G. Karpova et al. 2008. RNA interference of the clock gene period disrupts circadian rhythms in the cricket *Gryllus bimaculatus*. J Biol Rhythms. 23(4): 308–318.

Mutti, N.S., Y. Park, J.C. Reese et al. 2006. RNAi knockdown of a salivary transcript leading to lethality in the pea aphid, *Acyrthosiphon pisum*. J Ins Sci. 6: 1–7.

Niu, B.L., W.F. Shen, Y. Liu et al. 2008. Cloning and RNAi-mediated functional characterization of *MaLac2* of the pine sawyer, *Monochamus alternatus*. Insect Mol Biol. 17(3): 303–312.

Nunes, C.C. and R.A. Dean. 2012. Host-induced gene silencing: a tool for understanding fungal host interaction and for developing novel disease control strategies. Mol Plant Pathol. 13: 519–529.

Oerke. 2006. Crop losses to pests. The J Agric Sci. 144(01): 31–43.

Pimentel, D. 2005. Environmental and economic costs of the application of pesticides primarily in the United States. Env Dev and Sustainability. 7: 229–252.

Pitino, M., A.D. Coleman and M.E. Maffei. 2011. Silencing of aphid genes by dsRNA feeding from plants. PLoS One. 6: e25709.

Price, D.R. and J.A. Gatehouse. 2008. RNAi-mediated crop protection against insects. Trends Biotechnol. 26(7): 393–400.

Pridgeon, J.W., L. Zhao, J.J. Becnel et al. 2008. Topically applied *AaeIAP1* double-stranded RNA kills female adults of *Aedes aegypti*. J Med Ent. 45: 414–420.

Pereira, C.A., G.D. Alonso, M.C. Paveto et al. 2000. *Trypanosoma cruzi* arginine kinase characterization and cloning. J Biol Chem. 275: 1495–1501.

Quan, G.X., T. Kanda and T. Tamura. 2002. Induction of the white egg 3 mutant phenotype by injection of the double-stranded RNA of the silkworm white gene. Insect Mol Biol. 11(3): 217–22.

Rajagopal, R., S. Sivakumar, N. Agrawal et al. 2002. Silencing of midgut aminopeptidase N of *Spodoptera litura* by double-stranded RNA establishes its role as *Bacillus thuringiensis* toxin receptor. J Biol Chem. 277(49): 46849–46851.

Rao, Z., W. He, L. Liu et al. 2012. Identification, expression and target gene analyses of Micro RNAs in *Spodoptera litura*. PLoS One. 7: e37730.

Roignant, J.Y., C. Carre, B. Mugat et al. 2003. Absence of transitive and systemic pathways allows cell-specific and isoform-specific RNAi in *Drosophila*. RNA. 9: 299–308.

Ruby, J.G., A. Stark, W.K. Johnston et al. 2007. Evolution, biogenesis, expression, and target predictions of a substantially expanded set of *Drosophila* microRNAs. Genome Res. 17: 1850–1864.

Saleh, M.C., R.P. van Rij, A. Hekele et al. 2006. The endocytic pathway mediates cell entry of dsRNA to induce RNAi silencing. Nat Cell Biol. 8: 793–802.

Sanahuja, G., R. Banakar, R.M. Twyman et al. 2011. *Bacillus thuringiensis*: A century of research, development and commercial applications. Plant Biotechnol J. 9: 283–300.

Schlüns, H. and R.H. Crozier. 2007. Relish regulates expression of antimicrobial peptide genes in the honeybee, *Apis mellifera*, shown by RNA interference. Insect Mol Biol. 16(6): 753–759.

Shih, J.D. and C.P. Hunter. 2011. *SID-1* is a dsRNA-selective dsRNA-gated channel. RNA. 17: 1057–1065.

Sidahmed, A.M. and B. Wilkie. 2010. Endogenous antiviral mechanisms of RNA interference: a comparative biology perspective. Methods Mol Biol. 623: 3–19.

Siomi, H. and M.C. Siomi. 2009. RISC hitches onto endosome trafficking. Nat Cell Biol. 11: 1049–1051.

Sumitani, M., D.S. Yamamoto, J.M. Lee et al. 2005. Isolation of white gene orthologue of the sawfly, *Athalia rosae* (*Hymenoptera*) and its functional analysis using RNA interference. Insect Biochem Mol Biol. 35(3): 231–240.

Szittya, G., D. Silhavy, A. Molnar et al. 2003. Low temperature inhibits RNA silencing-mediated defence by the control of siRNA generation. EMBO J. 22(3): 633–640.

Takahashi, T., A. Hamada, K. Miyawaki et al. 2009. Systemic RNA interference for the study of learning and memory in an insect. J Neurosci Methods. 179(1): 9–15.

Tang, T., C. Zhao, X. Feng et al. 2012. Knockdown of several components of cytochrome P450 enzyme systems by RNA interference enhances the susceptibility of *Helicoverpa armigera* to fenvalerate. Pest Manag Sci. 68: 1501–1511.

Tao, X.Y., X.Y. Xue, Y.P. Huang et al. 2012. Gossypol-enhanced P450 gene pool contributes to cotton bollworm tolerance to a pyrethroid insecticide. Mol Ecol. 21: 4371–4385.

Terenius, O., A. Papanicolaou, J.S. Garbutt et al. 2011. RNA interference in *Lepidoptera*: An overview of successful and unsuccessful studies and implications for experimental design. J Insect Physiol. 57: 231–245.

Tian, H., H. Pen, Q. Yao et al. 2009. Developmental control of a Lepidopteran pest *Spodoptera exigua* by ingestion of bacteria expressing dsRNA of a non-midgut gene. PLoS One. 4: 1–13.

Timmons, L. and A. Fire. 1998. Specific interference by ingested dsRNA. Nature. 395(6705): 854.

Tomoyasu, Y. and R.E. Denell. 2004. Larval RNAi in *Tribolium* (*Coleoptera*) for analyzing adult development. Dev Genes and Evolution. 214: 575–578.

Turner, C.T., M.W. Davy, R.M. MacDiarmid et al. 2006. RNA interference in the light brown apple moth, *Epiphyas postvittana* (Walker) induced by double-stranded RNA feeding. Insect Mol Biol. 15(3): 383–391.

Ulvila, J., M. Parikka, A. Kleino et al. 2006. Double-stranded RNA is internalized by scavenger receptor-mediated endocytosis in *Drosophila* S2 cells. J Biol Chemistry. 281: 14370–14375.

van der Krol, A.R., L.A. Mur, M. Beld et al. 1990. Flavonoid genes in petunia: addition of a limited number of genes copies may lead to a suppression of gene expression. Plant Cell. 2: 291–299.

Vermehren, A., S. Qazi and B.A. Trimmer. 2001. The nicotinic alpha subunit MARA1 is necessary for cholinergic evoked calcium transients in Manduca neurons. Neurosci Lett. 313(3): 113–6.

Voinnet, O. 2005. Non-cell autonomous RNA silencing. FEBS Lett. 579: 5858–5871.

Walshe, D.P., S.M. Lehane, M.J. Lehane et al. 2009. Prolonged gene knockdown in the tsetse fly Glossina by feeding double stranded RNA. Insect Mol Biol. 18(1): 11–19.

Wang, Z., Y. Dong, N. Desneux et al. 2013. RNAi silencing of the *HaHMG-CoA reductase* gene inhibits oviposition in the *Helicoverpa armigera* cotton bollworm. PLoS One. 8(7): e67732.

Wang, Y., H. Zhang, H. Li et al. 2011. Second-generation sequencing supply an effective way to screen RNAi targets in large scale for potential application in pest insect control. PLoS One. 6: e18644.

Whangbo, J.S. and C.P. Hunter. 2008. Environmental RNA interference. Trends Genet. 24(6): 297–305.

Whyard, S., A.D. Singh and S. Wong. 2009. Ingested double-stranded RNAs can act as species-specific insecticides. Insect Biochem Mol Biol. 39: 824–832.

Winston, W.M., C. Molodowitch and C.P. Hunter. 2002. Systemic RNAi in *C. elegans* requires the putative transmembrane protein SID-1. Science. 295: 2456–2459.

Winston, W.M., M. Sutherlin, A.J. Wright et al. 2007. *Caenorhabditis elegans* SID-2 is required for environmental RNA interference. Proc Natl Acad Sci USA. 104: 10565–10570.

Winter, J., S. Jung, S. Keller et al. 2009. Many roads to maturity: microRNA biogenesis pathways and their regulation. Nat Cell Biol. 11: 228–234.

Wu, Q.Y., F. Li, W.J. Zhu. et al. 2007. Cloning, expression, purification, and characterization of *arginine kinase* from *Locusta migratoria manilensis*. Comp Biochem Physiol Part B. 148: 355–362.

Xie, Q. and H.S. Guo. 2006. Systemic antiviral silencing in plants. Virus Res. 118: 1–6.

Xiong, Y., H. Zeng, Y. Zhang et al. 2013. Silencing the *HaHR3* gene by transgenic plant-mediated RNAi to disrupt *Helicoverpa armigera* development. Int J Biol Sci. 9: 370–381.

Xue, X.Y., Y.B. Mao, X.Y. Tao et al. 2012. New approaches to agricultural insect pest control based on RNA interference. Adv in Insect Physiol. 42: 73–117.

Yang, Y., Y. Jittayasothorn, D. Chronis et al. 2013. Molecular characteristics and efficacy of *16D10* siRNAs in inhibiting root-knot nematode infection in transgenic grape hairy roots. PLoS One. 8: e69463.

Youssef, R.M., K.H. Kim, S.A. Haroon et al. 2013. Post-transcriptional gene silencing of the gene encoding *aldolase* from soybean cyst nematode by transformed soybean roots. Exp Parasitol. 134: 266–274.

Yu, F., H. Yao, P. Zhu et al. 2007. Let-7 regulates self renewal and tumorigenicity of breast cancer cells. Cell. 131: 1109–1123.

Zha, W., X. Peng, R. Chen et al. 2011. Knockdown of midgut genes by dsRNA-transgenic plant-mediated RNA interference in the hemipteran insect *Nilaparvata lugens*. PLoS One. 6: e20504.

Zhang, H., H.C. Li and X.X. Miao. 2013. Feasibility, limitation and possible solutions of RNAi-based technology for insect pest control. Insect Sci. 20: 15–30.

Zhang, X., J. Zhang and K.Y. Zhu. 2010. *Chitosan*/double-stranded RNA nanoparticle mediated RNA interference to silence *chitin synthase* genes through larval feeding in the African malaria mosquito (*Anopheles gambiae*). Insect Mol Biol. 19: 683–693.

Zhao, Y., G. Yang, G. Wang Pruski et al. 2008. *Phyllotreta striolata* (Coleoptera: *Chrysomelidae*): *Arginine kinase* cloning and RNAi-based pest control. European Journal of Entomol. 105(5): 815–812.

Zhou, X., M.M. Wheeler, F.M. Oi et al. 2008. RNA interference in the termite *Reticulitermes flavipes* through ingestion of double-stranded RNA. Insect Biochem Mol Biol. 38(8): 805–815.

Zhu, J.Q., S. Liu, Y. Ma et al. 2012. Improvement of pest resistance in transgenic tobacco plants expressing dsRNA of an insect-associated gene EcR. PLoS One. 7: e38572.

Zhu, Q., Y. Arakane, R.W. Beeman et al. 2008. Functional specialization among insect chitinase family genes revealed by RNA interference. Pro Natl Aca Sci USA. 105(18): 6650–6655.

CHAPTER 9

Overview of the Biosafety and Risk Assessment Steps for Insect-resistant Biotech Crops

Venera Kamburova and *Ibrokhim Y. Abdurakhmonov**

INTRODUCTION

One of the significant global challenges faced by agriculture is the risk of damage to crops by pests. According to Ross and Lembi (1985), of the estimated 67000 pest species that damage agricultural crops, approximately 9000 species are insects and mites. The resulting spread of such plant pests and pathogens leads to not only crop losses but also a reduction of crop quality. The estimated crop losses from herbivores differ in various cultivars and specifically the losses are 52% in wheat, 58% in soybean, 59% in maize, 74% in potato, 83% in rice and 84% in cotton (Kumar et al. 2008). The total worldwide agricultural loss from pest and pathogens is estimated at $ 1.4 trillion (and it is equal to 5% of global gross domestic product; FAO 2013, 2015). Traditionally, pesticides were widely used for pest control. However, the wide use of pesticides causes great harm to both human health and the environment. Besides, pesticide application essentially results in increased costs of crop production (Kumar et al. 2008).

The more recent applications in crop research relying on biotechnological methods of creation of novel plant crops with herbivore resistant traits resulted in a drastic reduction of pesticide use, thus decreasing their negative effects on human health and the environment (James 2007; Kumar et al. 2008). The widespread planting of biotech

Center of Genomics and Bioinformatics, Academy of Sciences of Uzbekistan, University street-2, Kibray region, Tashkent 111215, Republic of Uzbekistan.
Email: venera.kamburova@genomics.uz
* Corresponding author: geneomics@uzsci.net; ibrokhim.abdurakhmonov@genomics.uz

crops resulted in huge economic benefits to farmers because of the high productivity and low costs of agricultural production (DeFrancesco 2013; James 2015). Currently, genetically modified (GM) plants including those with resistance to pests are widely used in agriculture. In 2015, GM crops were planted on nearly 180 million hectares of agricultural land, out of which 205 GM crops facilitated pest management (James 2015). However, despite the enormous potential benefits of modern biotechnology, including insect-resistant (IR) crops (Pray et al. 2002), it is believed that the products of modern biotechnology can have potential hazards (real or perceived) to human health or environment. Consequently, current crop biotechnology research pays significant attention to the development of evidence-based approaches to assess the potential risks of the application of genetically modified organisms (GMOs) to ensure adequate protection of human health and environment. Among others, the potential risks to human health associated with the use of GMOs that were examined (Bergmans et al. 2008; Keese 2008; Goodman and Tetteh 2011; Roberts et al. 2015), include (1) toxicity and/or allergenicity GM IR crops; (2) changes in individual gene activity under the effect of foreign DNA insertion that affect consumer diet derived from these GMOs; (3) horizontal transfer of transgenes to other organisms, such as antibiotic resistance marker genes from GM IR crops to micro-organisms of the digestive tract and (4) non-target effect of the introduced constructions on the other organisms, which can potentially lead to the suppression of non-target genes (Table 1).

It should be noted that the above risk groups are differently correlated with various types of IR GM crops. Thus, transgenic organisms (organisms derived through insertion of novel genes, which are extrinsic to these species) are characterized by the following risk types: toxicity and/or allergenicity, unintended effects of genetic modification and consequently the deterioration of consumer properties, as well as horizontal gene transfer (HGT) to other organisms. While for cisgenic organisms (organisms derived due to manipulation with its own gene or genes from related species), the risks are associated with toxicity, allergenicity and pleiotropic genetic effects and characterized less than the risks of non-target effects and the horizontal marker genes transfer.

Despite the difference in nature of risks for the cis- and transgenic IR crops, the strategy of their safety assessment is based on the general principle "substantial equivalence" for all GM IR crops, which was developed by the Organization for Economic Co-operation and Development OECD (1993, 2000). The information characterizing the initial organism, that is the gene source for transgenesis and the features of genetic modification, are the subject for scrutiny. These checks are carried out for the identification of non-analog specifics in the novel products and

Table 1. Analyses of studies on biosafety of insect-resistant GM plants.

No.	Main method of insect resistant GM plant creation	Assessed parameter	References
1	**Traditional transgenesis using of several genes with insecticidal properties**	Toxicity	Li and Romeis (2010) Wang et al. (2013) Li et al. (2014) Baktavachalam et al. (2015) Guo et al. (2015) Mezzomo et al. (2016)
		Allergenicity	Creighton et al. (1993) Bucchini and Goldman (2002) Randhawa et al. (2011) Hammond and Jez (2011) Baktavachalam et al. (2015) Rubio-Infante and Moreno-Fierros (2016)
		Unintended effects of genetic modification	Baktavachalam et al. (2015)
		Horizontal gene transfer probability	Kleter et al. (2005) Baktavachalam et al. (2015) Koch et al. (2015)
2	**RNA interference**	Non-target genes suppression	Espinoza et al. (2013) Petrick et al. (2013) Telem et al. (2013) Casacuberta et al. (2014) Ramon et al. (2014) Roberts et al. (2015) Xiong et al. (2015) Perkin et al. (2016)
3	**Genome editing technologies (TALEN, CRISPR)**	Non-target effects in gene editing	Mussolino et al. (2011) Gaj et al. (2013) Graham and Root (2015) Sprink et al. (2016) Hartley et al. (2016) Perkin et al. (2016)

in the original non-modified food/feed products, as well as a nutritional value of food products (Kohl et al. 2015).

Further, a comparative analysis between GM IR organisms and their original (unmodified) counterparts is conducted. Agronomic features, inserted gene products, a composition of the key chemical components (including nutritional and anti-nutritional), a profile of major metabolites, and effects of feedstock processing are compared for the safety and assessment the possible risks (FAO/WHO 2000, 2001a; Hong et al. 2014).

According to the analysis results received within a substantial equivalence study, the novel product could be (FAO/WHO 2000, 2001a; Hong et al. 2014): (1) equivalent to the non-modified analog by the selected essential characteristics; (2) equivalent to the non-modified analog except one (or several) substantial(s), or well-defined feature(s); (3) non-equivalent to the non-modified analog by the essential features. In the first case, if a new product is a substantially equivalent by composition and nutritional value compared to existing analogs with a safe usage history, then, it is considered as safe and does not require a thorough safety assessment (OECD 1993; FDA 1992; Kuiper et al. 2001; Maryanski 1995). The other two cases, however, require a careful assessment of the safety of the cis- and transgenic organisms (Roberts et al. 2015; FAO 2000, 2001a, 2001b; Royal Society 2002).

To illustrate the differences in the strategy of risk assessment for the cis and transgenic organisms, it's worth considering the assessments in terms of the new generation of genome modification tools such as RNAi.

Risks Associated with Transgenic Insect Resistant Crops

Despite the existence of different strategies for the engineering of insect-resistant (IR) transgenic crops (utilizing several genes with insecticidal properties such as inhibitors of insect digestive proteases, α-amylase, lectin and others), a high percentage of the transgenic IR crops in cultivation worldwide are based on the insertion of *cry* genes encoding *Bacillus thuringiensis* (or Bt) toxin in genome of host plant (Fontes et al. 2002). According to the International Service for the Acquisition of Agri-biotech Applications (ISAAA) data at the end of 2014, an estimated 27.4 million hectares of land were planted with crops containing the Bt gene (http://www.isaaa.org/kc).

The global commercial use of Bt-crops is based on its specific mode of action involving the Bt (or Cry) toxins' specific activities against insect species of the orders *Lepidoptera* (moths and butterflies), *Diptera* (flies and mosquitoes), *Coleoptera* (beetles), *Hymenoptera* (wasps, bees, ants and sawflies) and nematodes (Bravo et al. 2007; Palma et al. 2014). Despite the fact that Bt toxins have seen extensive commercial usage, the specificity of their mode of action are still controversial (Tabashnik et al. 2015). The toxic effect of Bt-protein involves a multi-stage process consisting the following phases: ingestion by susceptible insects, solubilizing, and the activation of protoxin to toxin in the insect nymph digestive fluid. The toxin enters the peritrophic matrix and binds to cadherins (specific receptors on the gut cells membrane). Further events can be depicted through two different scenarios. According to the first scenario, the binding of toxin to cadherin activates cell death pathway (Tabashnik et al. 2015). The second scenario suggests the formation of toxin oligomers that bind to GPI-anchored proteins, which get concentrated on regions of the cell membrane called lipid rafts. Accumulation of toxin oligomers results in toxin insertion in the membrane, pore formation, osmotic cell shock, and ultimate insect death (Bravo et al. 2007; Palma et al. 2014; Koch et al. 2015; Tabashnik et al. 2015).

It has been shown that *CRY* proteins are safe for vertebrates (Siegel 2001; Federici and Siegel 2008; Gatehouse et al. 2011). However, despite the fact that there is no evidence for Bt-protein toxicity on mammalian systems (US EPA 2001), all Bt-based crops like other transgenic GM must be subjected to the risk assessment procedures.

The first step in testing of substantial equivalence of a Bt transgenic organism is to identify the differences between the novel transgenic cultivar and its analogue *CRY* proteins that are expressed. The subsequent parameters that are evaluated when the Bt transgenic cultivar goes through the complete procedure of safety assessment (Wang et al. 2015) include: potential toxicity, potential allergenicity, the possibility of transfer of antibiotic resistance genes to microorganisms of the digestive tract, the probability of a potential deterioration of the nutritional value and nutrients assimilation (OECD; FAO 2015).

Toxicity Assessment

A common strategy for the evaluation of the potential toxicity of new IR trangenic crops (FAO 2003, 2015) first ensures that the introduced and expressed transgenic product test results do not differ from a known component of plant foods that have a long history of safe use. In such cases, the related toxicity studies of new products are not conducted further. In other cases, the following analyses is mandated (FAO 2003, 2015): (1) determination of potential toxins concentration in the edible parts of plants; (2) an establishment of specific weight of the transgenic product in the diet of certain identified population groups; (3) a comparison of amino acid sequences (of the expressed transgenic proteins) with known toxins and food antagonists by electronic databases; (4) a stability analysis of the new substances towards heat treatment; (5) a determination of the destruction rate of potential toxins in the gastrointestinal tract (in model systems); (6) an analysis of the toxicity level of new substances in model systems (cell culture *in vitro*); and (7) toxicity analysis in experiments based on the forced feeding of the laboratory or pet food mixed with transgenic products for extended time periods (chronic experiment) or for monitored short times using high concentrations of the transgenic products (acute experiment) (WHO 1995, 2000). Further, an analysis of toxicity level *in vitro* is carried out, if preliminary tests revealed that: (1) product level of evaluated proteins in GM IR crops is significantly higher than the product level of similar natural proteins; (2) modified protein showed toxicity and if they are natural agents with antibiological functions; (3) new protein had no history of safe use as food; and (4) new protein has a high level of the novel proteins stability to physical and chemical degradation (WHO 1995, 2000).

Further toxicity research should be conducted individually depending on the specific properties of modified protein. The general *in vivo* tests include: toxicity determination in acute, subchronic, and chronic experiments by feeding the purified protein to laboratory animals with the definition of the common indicators such as half lethal dose (LD_{50}), acceptable daily intake (ADI), and the no-observed-adverse-effect-level (NOAEL) (FAO 2015). It should be noted that chronic tests are rarely used because the known toxic proteins usually act via acute mechanisms at low doses, and chronic toxicity of any protein are not proved (Maryanski 1995).

The toxic effects of Bt-proteins are determined by various factors in *in vivo* experiments such as mortality level, the dynamics of body weight gain, weight change of individual organs, as well as measuring other, more specific parameters (e.g., the level of cell proliferation of certain tissues, level of immune protection, etc.). Furthermore, the food safety assessments also used a number of other narrowly focused tests such as the analysis of assessed binding protein with cells receptors of the mammalian gastrointestinal tract, analysis of the hemolytic potential and immunotoxicity of the transgenic products (Kuiper et al. 2001).

It should be noted here that the risk assessment of Bt-based transgenic crops, carried out within the framework of a scientific process-based integrated approach, thus far has provided solid evidence that diverse GM plants engineered with Bt-proteins are completely safe for human consumption and animal health (US EPA 2001; Li and Romeis 2010; Li et al. 2014; Baktavachalam et al. 2015). Specifically, all studies on risk assessment of Bt-based GM crops using animal models reported that in mouse acute oral feeding studies (in which doses of *Cry*-proteins as high as 5,000 mg/kg body weight are used), no biohazardous or any kind of adverse effects were observed (Sjoblad 1992; US EPA 2001; Baktavachalam et al. 2015; Mezzomo et al. 2016). Further, investigations of long term toxicity of Bt-based GM crops (such as maize and rice) also reported that such cultivars are completely non-toxic towards model animals and have no negative influence on the main organ systems of mammalian species (Betz et al. 2000; Wang et al. 2013; Guo et al. 2015; Baktavachalam et al. 2015).

Allergenicity Assessment

There exist general procedures (set protocols and sequences of analytical experiments) developed by World Health Organization (WHO) and Food and Agriculture organization (FAO) experts to assess allergenic potential of diverse GM crops including insect-resistant transgenic plants (FAO 2001b, 2015). A procedure of risk assessment of allergenicity of GM plants and novel GM foods, including *Cry*-based crops, is based on tested and reliable protocols set forth by diverse safety agencies (FAO 2001, 2015; Bucchini and Goldman 2002; Baktavachalam et al. 2015). Firstly, the fact that expressed transgenic proteins may be completely foreign to the original organism proteome, they should be evaluated in terms of their potential allergenicity as related to human consumption. The expressed transgenic protein can be similar to any protein with known allergenic potential, and/or proteins that previously were not used in food with unknown allergenic characteristics. Secondly, the necessity of risk assessment stems from the potential of the increase of host organism allergenicity due to unintended effects of genetic modifications. Thirdly, it comes from the fact that there is no single universal reliable test for assessment of the general antigen-allergenicity level (Bernstein et al. 2003). Therefore, every risk assessment procedure should provide an integrated, consistent approach, while considering the specifics of every case separately. The final investigation of the risk assessment should answer the following questions: is the newly synthesized protein an allergen, and has it changed the allergenic potential of the GMOs and related products due to unintentional modification effects? (Goodman 2013; FAO 2015).

The literature review for the characteristics of currently grown GM plants, including Bt-based transgenic crops, revealed that among the expressed gene products that are integrated in the original genome, there are practically no proteins and metabolites that would have the potential toxicity or allergenicity in relation to endothermic organisms (Rubio-Infante and Moreno-Fierros 2016). This is due to the fact that so far only the genes encoding the formation of enzymes, that are similar between plants and microorganisms, are included in engineering of transgenic organisms (Batista 2005; Randhawa et al. 2011).

Among all transgenic cultivars developed, a very limited number of transgenic plants were developed as a result of genetic modification with new compounds not typical to the genome of conventional varieties of species, and Bt-transgenic cultivars belong to this category of GM crops. However, it should be noted that such atypical substances are thoroughly tested. In allergenicity testing, as illustrated by Cry-proteins, the following physicochemical and biological characteristics should be considered (FAO 2001, 2015): (1) they are unstable compounds, easily denatured even at relatively low temperatures and pH medium changes; (2) they are rapidly cleaved under the proteases action into amino acids (not toxic and/or allergic normally) in gastrointestinal tract and that, (3) their content is very low in plant tissues (Bucchini and Goldman 2002; Bernstein et al. 2003; Baktavachalam et al. 2015). All these factors have suggested a low probability for *Cry*-proteins to cause allergic reactions because most of the known allergenic proteins are stable to digestion in the gastrointestinal tract and to various types of processing (including thermal stability). A molecular weight of 10–70 kDa, and a quantity of more than 1% food content are crucial characteristics for substances to be classified as allergens (Yerger et al. 1990; Creighton 1993).

Despite the absence of a single universal test for an accurate estimation of allergenicity level risk, every developed protocol begins with a description of the allergic potential of transgenic source, specifically, the potential allergenicity of the donor (FAO 2001; Bernstein et al. 2003). Such a description could establish that the genetically engineered protein (transgene product) may not cause allergic reactions at a consumption level and with a high probability will not be the cause of any allergic reaction at the expression level in a transgenic organism. Another possibility may be that an expression of the transgenic protein in the donor organism genome could likely increase the allergenic potential of consumer (Nordlee et al. 1996). Based on these hypotheses, the first stage of risk assessment seeks to establish whether the transgene source is recognized as a major or minor allergen, or if it is not a known allergen. If the transgene source belongs to a major or minor allergenic sources, the final GM crops and its food products are recognized as allergenic unless otherwise proven (FAO 2015).

After allergenic potential of donor GM organism has been established, the next step of the adopted procedure is the comparison of amino acid sequence of the novel protein with the amino acid sequence of known allergens, present in the special bioinformatics computer databases (e.g., Swiss Prot, TrEMBL, GenBank, PIR) (Randhawa et al. 2011). The purpose of comparing the amino acid sequence is to establish the structural similarity of newly synthesized protein to known allergens. A structural similarity for establishing allergenicity potential is a minimum of 35% of identified sequences of random fragments of at least 80 amino acids detected or the full identity of six contiguous amino acids in the compared proteins (probable minimum linear epitope) (FAO 2001). Besides the comparative analysis of the amino acid sequence, a physicochemical test towards protein stability to gastrointestinal tract proteases is conducted at the early stages of research (FAO 2001).

The above mentioned preliminary indirect allergenicity tests aggregately allow evaluating a certain probability towards establishing whether the estimated protein is an allergen. A positive result indicates a high probability that the tested proteins may have allergenic potential. A negative result of indirect tests, however, can still not be an absolute proof that the tested proteins have no allergenic potential. Therefore, after characteristics of preliminary structural and physicochemical features of the studied proteins, a risk assessment procedure usually continues towards additional steps (FAO 2015).

The subsequent steps provide the process of conducting specific immunological studies that seek to conclusively establish whether the tested proteins are allergens or not. For proteins that originated from known allergenic sources or those with structural homology of known allergens, the risk assessment procedure recommends conducting the so-called specific serum screening. Radioallergosorbent test (RAST) and enzyme-linked immunosorbent assay (ELISA) are solid-phase quantitative immunoassays that are the most commonly used diagnostic tests, applied for *in vitro* immunological studies (IFIC 2001). A positive result of target serum screening indicates that the transgenic protein is most likely an allergen. A negative result combined with a protease-resistance absence and a lack of homology with known allergens indicates that the tested protein is not likely an allergen. It should be noted here that the result of *in vitro* immunological test might be sufficient to conclude the absence of allergenic potential in the transgene product. However, under special circumstances, there exists a mandate for conducting additional immunological studies *in vivo* in animal models or in clinical trials. In the clinical analysis, *in vivo* skin test (SPT—skin prick test) is commonly used (OECD 1997) for allergenicity test of the transgenic proteins. Double-blind placebo-controlled food challenge (DBPCFC) is a more sensitive immunological test that is the most accurate method to confirm the substance allergenicity and can be applied at the final stages of the risk assessment (IFIC 2001).

The risk assessment should also consider the possibility that the genetic modification may increase the allergenic potential of the original host organism as a result of unintentional (pleiotropic) effects (Cellini et al. 2004; Ladics et al. 2015). A preliminary analysis of substantial equivalence is crucial at risk assessment in such cases. Allergenicity assessment methods, adequate for detecting unintended effects of genetic modification (UEGM), should be applied in case of non-equivalence of GMO to its analog based on essential features. UEGM makes it possible to identify allergenic potential using *in vitro* solid-phase immunological assays and DBPCFC. Thus, for measuring changes in allergenic potential, an assessment of consumer organism for the allergic immunological reaction is measured in response to the proteins mixture rather than an individual transgene product (Cellini et al. 2004; Ladics et al. 2015; Koch et al. 2015).

Reports based on studies of Creighton et al. (1993), Randhawa et al. (2011), Hammond and Jez (2011), Baktavachalam et al. (2015), and others revealed that due to instability by heating and pH changes, *Cry* proteins cannot pose a serious allergen problem in humans. This fact is confirmed by Creighton et al. (1993) who showed that there exists no significant alignment and similarity of Cry proteins at the domain level with any of the known allergens that also reveals that there exists no potential risk of allergenic cross-reactivity. In addition, Baktavachalam et al. (2015) have shown that the product of *Cry*-genes used for engineering resistance to herbivores in crops has no history of causing allergy and due to absence of any amino acid sequence similarity with known allergenic proteins, denaturation on heating, readily digestible by pepsin, there is also a lower probability they would cause adverse health effects.

Assessment of Potential Nutritional Value Deterioration due to Unintended Effects

In the ideal scenario, plant genetic modification with linear event sequence initiated by transformation does not affect the activity of other genes, and the specific GM crops differs from its counterpart simply by expression of engineered transgene products (Rischer and Oksman-Caldentey 2006; Herman and Price 2013). In this regard, a possible risk assessment is focused only on the novel artificially modified gene(s). However, in practice, the linear model of genetic modification does not always describe the real processes occurring in living cells including complex interactions between its molecular and genetic components.

Due to the presence of such scientific uncertainty on a possible non-linearity of genetic modification, an additional feature that may occur is the manifestation of target features (intended effect) or a modification of pre-existing features (unintended effect) (Cellini et al. 2004). UEGM can manifest as suppression or changing of expression level of previously active genes or an expression activation of previously "silent" genes due to random insertion of DNA sequences into the plant genome. However, some parts UEGM might be partially predicted, while others are unpredictable and associated with a scientific uncertainty (Nakayachi 2013).

By a degree of impact, UEGM can be harmful, beneficial and neutral in regards to both the modified organism and human health or non-target organisms (NTOs). Safety analysis of transgenic organisms and new food products is aimed to minimize any unintended harmful effects due to genetic modification (Herman and Price 2013). Considering the nature of UEGM, the most predictable consequence is qualitative and/or quantitative change in IR GM crops composition compared to its counterpart wild-type.

Therefore, UEGM is identified as a crucial feature within the establishment of GM Bt-cultivars' substantial equivalence and not modified analogs by conducting a wide range of comparative biochemical studies (FAO 2001, 2001a, 2001b; Cellini et al. 2004; Ladics 2015).

The various strategic approaches that are applied for UEGM identification include both target and non-target approaches (Kuiper et al. 2001; Cellini et al. 2004; Ladics 2015). In the case of a targeted approach, a limited number of key chemical components of IR GM plants and new food products essential for human health could be studied, and whether the contents could be changed unintentionally as a result of genetic modification. The more the components (such as enzymes, metabolites) analyzed, the greater probability to find the unpredictable UEGM (Kuiper et al. 2001; Cellini et al. 2004). Such an approach is the most commonly used and is proved to be the most effective research method to analyze UEGM. The targeted UEGM analysis strategy is however, limited by possible appearance in the modified organism unknown toxic components or food antagonists that have no human consumption history, therefore cannot be subject to a target identification (Ladics 2015).

An alternative approach to UEGM determination is the so-called non-targeted approach. It uses specialized analytical methods that allow determining the gene expression status and detecting qualitative and quantitative changes in composition of numerous chemical compounds at the intracellular level. The order of gene expression in cells is assessed at transcript, proteome and metabolite levels (Ladics 2015). In this case, a profile analysis (finger-printing) can detect a full spectrum of certain classes of chemical compounds in cells and tissues, regardless of their biological function, and thereby, it gives individual characteristics of the studied genotype (Kok et al. 1998; Noteborn et al. 1998). This approach is quite time-consuming and rarely used, but in the long term, it effectively identifies many UEGM manifestations.

UEGM fingerprinting is carried out to analyze quantitative signs. Compared samples, grown under the same environmental conditions are exposed to assessment (Kok et al. 1998; Rischer and Oksman-Caldentey 2006). In this manner, an assessment generally involves simultaneous study of GMOs and their counterparts testing in different geographical and climatic conditions. In such comparative studies, the differences of GMO and its non-modified analogue in the key component molecular content and metabolites are authentically determined. Conventional statistical methods are used for data analysis and determining the difference and similarity of comparison groups at the required significance level (Kok et al. 1998).

Additionally, in the case of UEGM identification, further risk analysis is conducted to allow the assessment of damage probability to human health or NTOs as result of the manifestation of the identified UEGM. In such investigations, the possibility of toxic and/or increase of allergenic potential as well as a modification ability of their nutritional value are taken into account. The assessment of toxic and/or allergenic potential is conducted according the methodology proposed by WHO and FAO experts (FAO 2001, 2015). At the same time, a careful analysis of the potential UEGM is the most important steps for organism types that can be potentially hazardous for human health and NTOs. These include, potatoes and tomatoes (due to toxic glycoalkaloids synthesis), cotton (due to synthesis of gossypol) and other important crops (Herman and Price 2013).

Thus, it should be noted that in the current research involving the risk assessment of existing and novel Bt-based GM crops, the significant differences in the key components composition (caused by UEGM) are not accounted for (Baktavachalam et al. 2015). Searches in the PubMed database also revealed the absence of research publications that have any mention of whether the insertion of *cry* genes in plant genome deteriorates their nutritional value due to the unintended effects of the genetic modification. However, for reasons explained in this section, a complete exclusion of the possibility of harmful effects, as a result of UEGM seems impossible and further research should be encouraged (Chassy 2002).

Assessment of Horizontal Gene Transfer Probability

A crucial factor that should be considered as the cause for potential adverse effects of all types of GM cultivars to human health, NTOs and environment is a horizontal transfer of antibiotic resistance genes from GMO to digestive tract microflora (Keese 2008). The composition of any transgenic constructs (including those containing *cry* genes) includes together with a transgene and its regulatory elements, a marker gene, needed for the selection of transformed cells. Antibiotic resistance genes (e.g., kanamycin, ampicillin, streptomycin) are commonly used as marker genes. It is worth noting here that the use of these antibiotics in medical practice is now debated due to widespread resistance being developed by the microorganisms (Keese 2008).

Consequently, the expression of antibiotics resistance genes in GMOs cause serious concern and has been a subject of serious discussion (Ho 2014). The risk of these genes is associated primarily with the potential horizontal transfer of antibiotics resistance marker genes from GMO to human intestinal microflora (Keese 2008; Ho 2014). If horizontal gene transfer (HGT) occurs with relatively high frequency, it can have a negative impact on the effectiveness of conventional antibiotic therapy. A risk assessment of horizontal transfer marker genes on human health can certainly be carried within the framework of GMO risk assessment. The given requirement and the appropriate risk assessment criteria are included in the internationally accepted safety assessment process of GMOs and novel food products (EU 2001; Royal Society 2002).

The risk assessment of HGT is based on the following factors such as (1) the nature of microflora in the human gastrointestinal tract; (2) the potential toxicity and allergenicity of the marker gene products; and (3) the potential adverse effects caused by horizontal transfer of marker genes (Townsend et al. 2012; Nielsen et al. 2014).

Several mechanisms of horizontal gene transfer are known in nature, which can provide manifestations of novel traits in the recipient organism (Smalla et al. 2000; DeVries et al. 2001; Moens and Collard 2002; Keese 2008; Ruzzini and Clardy 2016). Among the known mechanisms, the processes of conjugation and transduction play a significant role in the exchange of genetic information between prokaryotes. A conjugative gene transfer *in vivo* (along with a transduction) is carried out at both intraspecies and interspecies levels. However, another genetic transformation mechanism, the so-called natural transformation, is more important for HGT risk assessment by higher GMO consuming, wherein an active transport of free extracellular DNA in the cytoplasm of bacterial cells is possible (Nielsen et al. 1998; Smalla et al. 2000; Keese 2008). Single-stranded DNA fragment capturing by bacterial cell that can theoretically be integrated into the bacterial genome as a result of homologous recombination, or by the formation of autonomous replication element has been discussed (Keese 2008; Townsend et al. 2012; Nielsen et al. 2014). The potential of natural transformation and insertion in the bacteria genome in case of the antibiotic resistance genes has led to the study of this process in terms of GM plant risk assessment. A number of international meetings submitted the results of these studies and recommendations to the appropriate risk assessment procedure (Townsend et al. 2012; Rizzi et al. 2012; Nielsen et al. 2014).

However, the result of a thorough expert study at the level of biosafety problems concluded that horizontal transfer of antibiotic resistance genes from GM crops and/or food to bacteria of the human intestinal microflora is an unlikely process (Kleter et al. 2005). Data obtained by diverse studies (Kleter et al. 2005; Baktavachalam et al. 2015; Koch et al. 2015) and others suggest that transfer of a gene from GE plants to intestinal microflora is improbable. Plant DNA traversing the gastrointestinal tract and tolerating digestive enzymes, while maintaining the original coding information, is a highly unlikely event. Further, even in a rare occurrence, HGT of cry genes from GE crops to microbes is unlikely to cause pathogenicity in receiving microbes residing in humans and animals. All the above facts indicate that horizontal transfer of relevant marker genes, widely used in creating GMOs is negligible. However, this risk is necessarily assessed by the GMO release, the character of required relevant information is provided, in particular, in Annex III of Directive 2001/18/EC (EU 2001). It can no longer be ignored that the potential HGT long-term effects and

HGT stability effects to novel, intensive usage of medical drugs. Therefore, the FAO/WHO experts suggest the avoidance of the use of those marker genes during process of GM crop engineering.

Research is now focused on methods of plant genetic modification without including the antibiotic resistance genes in transgenic constructs. An alternative to antibiotic resistance marker genes usage existed in the form of genes responsible for the synthesis of tryptophan decarboxylase, β-glucuronidase and several other enzymes (Kuiper et al. 2001). As a marker gene, a gene responsible for the synthesis of green fluorescent protein (GFP) was also used (Stewart et al. 2000). In addition, the methods of genetic transformation that allow to exclude the marker genes from the GMO at the stage after transformant selection are also in practice (Xiong et al. 2015). Furthermore, to reduce the risk of antibiotic markers, the methods of the removal of selective genes in transformants were optimized after selection procedures or obtaining of markerless transgenic lines using co-transformation, followed negative selection by selective genes in backcross generation (Breyer et al. 2014; Woo et al. 2015). Other possible methods of target removing of undesirable marker genes utilize the modern next generation technology of genome editing—CRISPR/Cas9, TALENs, and Zinc finger nucleases (Xiong et al. 2015; Khatodia et al. 2016; Luo et al. 2016; Puchta 2016). All these methodological approaches allow researchers to eliminate the risk of horizontal transfer of antibiotic resistance genes, however unlikely.

Thus, summarizing the analyzed information above, it can be concluded that, although the genes, not specific to a species, are inserted into the recipient organism genome during the development of transgenic organisms, the resulting GM crops (including insect-resistant transgenics) and their products do not have any specific radically novel risks to human health, NTOs and environment. The above described possible risks associated with genetic modification, are not typical for all newly created GM cultivars including insect-resistant varieties. As practice shows, deviation from the norm or from "expected" phenotype were very rare. These abnormalities can be detected visually more often and culled at the earliest testing steps. It can be concluded that the use of diverse procedures of risk assessment as applications of basic biosafety principles in practice can be viewed as effective precautionary measures.

Risk Assessment for GMOs Obtained using new Generation Genome Modification and Editing Tools

Current insect-resistant genetically modified (GM) crops are well adapted for modern crop production. However, these technologies are up against a crucial challenge to accommodate the enhancing demands on increased crop production. Additional pest management tools need to keep up with future agricultural demands, and RNA interference (RNAi)-based insect-resistant GM crops was one significantly effective response to this problem (Baum et al. 2007; Mao et al. 2011; Zha et al. 2011; Abdurakhmonov 2016).

RNAi is being widely used for insect control, both as a complementary practice to traditionally applied insecticidal application as well as production of GM plants (Abdurakhmonov 2016). RNAi corn plants with resistance to the western corn rootworm (*Diabrotica virgifera;* Coleoptera: Chrysomelidae) were developed by Baum et al. (2007). The resultant plants increased pest mortality and larval stunting by reducing translation of vacuolar H^+-ATPase subunit A (v-ATPase A) in the pest and further, plants suffered less root damage. Zha et al. (2011) transformed rice by RNAi constructs of several genes of *Nilaparvata lugens* (Hemiptera: Delphacidae). Although gene expression was suppressed, the insects were not killed by feeding on the GM rice. In a different approach to pest management, Mao et al. (2011) transformed cotton using RNAi of the P450 gene *CYP6AE14* in cotton bollworms (*Helicoverpa armigera;* Lepidoptera: Noctuidae). As a result, GM cotton suffered less damage and the larvae had reduced growth (Lundgren and Duan 2013).

Risk Assessment Steps for RNAi

The examples mentioned earlier illustrate that RNAi is one of the most advanced and safe biotechnological approaches which can be used for creation of new crops including insect-resistant transgenics. Although European Food Safety Association (EFSA) confirmed that RNAi crops are safe for human health and recognized the technology as a safer approach comparable to traditional breeding (EFSA 2012a, 2012b), currently RNAi insect-resistant crops (along with transgenic) are regarded as other genetic engineering products that mandate risk assessment. This is due to the fact that RNAi constructs can affect not only the targeted genes but also the native genes with significant homology to the RNAi fragments. This results in non-targeted gene suppression with possible harmful effects on human and animal health or on the environment, including NTOs (Casacuberta et al. 2014; Ramon et al. 2014; Roberts et al. 2015). This risk factor was shown to apply in diverse RNAi-based insect-resistant GM crops (Espinoza et al. 2013; Petrick et al. 2013; Telem et al. 2013; Roberts et al. 2015; Xiong et al. 2015). The non-target gene suppression can occur with genes that have a high degree of sequence similarity with presumed target gene, especially between a "seed region" of small interfering RNA (siRNA) (2–8 nucleotides thread guide), and a 3′ untranslated region of non-target gene (Jackson 2006).

However, in case of RNAi-based insect-resistant GM crops, in assessing substantial equivalence, the differences between the novel genetically modified organism and its analog is rarely detected (EFSA 2014). The risks related to toxicity, allergenicity and pleiotropic effect of genes are characteristic of IR GMOs to a lesser extent than risks of non-targeted effects and horizontal transfer of marker genes (Petrick et al. 2013; EFSA 2014; Ramon et al. 2014). Therefore, after substantial equivalence assessment of RNAi organisms, the probability of non-target effects of introduced RNAi construct using bioinformatic approaches was assessed (Roberts et al. 2015). A risk assessment, occurring as a result of horizontal marker-genes transfer is similar to the procedure, stipulated for transgenic organisms and discussed in detail in the previous section. In this section, we describe in detail the risk assessment of non-target impact introduced constructs in other organisms that could potentially lead to the suppressing of non-target genes.

To identify areas where more investigations may be useful to inform future risk assessments of RNAi-based insect-resistant GM crops, we first need to see beyond the possible negative impacts where the non-target organisms may be exposed to double-stranded RNA (dsRNA) from the RNAi-based crops, leading to adverse effects on humans (Raybould 2006; Wolt et al. 2010; Gray 2012; Roberts et al. 2015). The pathways of negative influence explain how RNAi-based IR GM crops could lead to adverse effects on NTOs through a chain of events with regard to both risk and impact. In addition, to conceptualize the relationship between the plant and NTOs, it is also useful to determine which steps can be most easily understood by testing. Firstly, it should be ensured that the plants must express a dsRNA. Subsequently, the NTOs must be affected by that dsRNA. After consumption, dsRNA should be resistant to degradation in the intestines. NTOs should assimilate sufficient amounts of dsRNA to activate endogenous RNAi mechanism. This may occur either locally at the point of capture (e.g., in cells lining the intestine), or systemically if NTOs are capable of causing systemic RNAi (Smagghe and Swevers 2014; Ivashuta et al. 2015). After activation, endogenous RNAi mechanism should lead to the degradation or silencing of a corresponding mRNA (Whyard et al. 2009; Zotti and Smagghe 2015). Finally, the loss of the transcript should have adverse effects on NTOs (Bachman et al. 2013). If any of these steps are deemed unlikely or impossible, the risk of non-targeted effects of RNAi-based insect-resistant GM crops may be considered negligible.

The assessment of possible risks of RNAi-based IR cultivars are challenged by the phenomena of pre- and post-RNAi development. In the preliminary development of RNAi-based IR crops, an important step is to design a dsRNA fragment for insertion into a hairpin structure. For this purpose, it is necessary to carry out a careful *in silico* analysis for detection of all possible types small interfering RNA (siRNA), and also their estimated targets in genomes of key recipients with use of target retrieval algorithms from available genomic databases. Despite existence of some restrictions connected with lack of the genomic sequences for off-target organisms (Casacuberta et al. 2014), this step can help to identify the first degree of risks which can be caused by the unintended functions of possible dsRNA from the gene of interest. If such unintended influence or highly specific off-target compliance of a consumer genome is revealed, then it is necessary to stop development of RNAi or change and optimize dsRNA for elimination of unintended effects. If dsRNA influences only the required function, then the subsequent step consists of the development of RNAi-based insect-resistant crops and the beginning of post-RNAi testing (Abdurakhmonov et al. 2016).

According to the scheme offered by Heinemann et al. (2013), the strategy includes the following steps of risks assessment: (1) assessment of RNAi-based insect-resistant varieties on manifestation of a required trait; (2) experimental quantitative determination of addressing the suppression at the gene expression level; and (3) a comparative sequencing of siRNA profile before and after a RNA interference. Although, at present, (4) there is no established regulatory base for food safety assessment, comparative research proteomic and metabolomic profiles of RNA-based insect-resistant varieties/product can be conducted (Ricroch et al. 2011; Clarke et al. 2013; Simó et al. 2014; Wang et al. 2015), and finally, (5) there is a need to carry out studies in cell/tissue culture and model animals to determine the safety of RNAi-based crops *in vivo*. These steps will allow determining and assessing the possible biological risks of RNAi-based crops.

Risk Assessment for new Generation Genome Editing Technologies

The generation of GM plants, including insect-resistant transgenics, using time-tested methods of genetic engineering is always followed by a combative debate connected with problems of their biosafety. The main claims of opponents of GM plants are associated with alien genes used during the genetic engineering process.

A similar process had to be in place in the analysis of genome editing technologies that requires the visualization of the operable gene expression and work of regulatory elements that help in expression of alien genes. The modern technologies of genome editing include ZFN systems (Zinc-finger nucleases), TALEN (Transcription Activator-Like Effector Nucleases) and CRISPR (Clustered Regulatory Interspaced Short Palindromic Repeats)/Cas9. These systems which were developed relatively recently are already proving to be effective and reliable tools of genomic engineering.

The roots of these technologies date back to 1996, when it has been shown that the protein "domain zinc fingers", connected to FokI-endonuclease domain, acts as the site-specific nuclease, cutting DNA in specific sites *in vitro* (Kim and Pabo 1998). Such a chimeric protein has modular structure as each domain "zinc fingers" affects one triplet of nucleotides (ZFN). This method has become a basis of editing the cultivated cells, including model plants (Townsend et al. 2009; Zhang et al. 2010). However, the technology based on ZFN has a number of limitations, including complexity and high cost of designing of protein domains for each specific genome locus, the probability of inexact cutting of target DNA due to one-nucleotide replacements or the wrong interaction between domains.

Therefore, an active search for newer methods of editing a genome were continued, and in recent years the search has led to creation of new technologies of editing genomes—the TALEN and CRISPR/cas systems (Xiong et al. 2015; Khatodia et al. 2016). These systems are unique in relative simplicity of design coupled with an outstanding performance in human, animal and plant cells. Such systems which are actively applied to various manipulations with genomes allow to solve complex problems, including those of mutant and transgene plants (Barrangou and Doudna 2016). Besides, chimeric proteins based on DNA-binding TALE domains and cas9 are widely used in experiments on regulation of transcription of genes, to study epigenetics and behavior of chromosomal loci in a cellular cycle (Xiong et al. 2015; Khatodia et al. 2016). Also, recently these technologies of genome editing were used in creation of new plant varieties with improved nutritious properties (Khatodia et al. 2016) and resistance to abiotic and biotic stresses (Khatodia et al. 2016; Lowder et al. 2016), including resistance to insects (Bachman et al. 2013; Shan et al. 2013; Champer et al. 2016; Perkin et al. 2016).

However, it should be noted that application of these technologies in generating insect-resistant crops also have their share of potential problems related to their biosafety (Sprink et al. 2016). As in the case of RNAi technology, the main challenge is the non-target effects in gene editing (Hartley et al. 2016; Perkin et al. 2016). Off-target effects in TALENs chimeric proteins system arise for several reasons: (1) it may be associated with difference in efficiency of binding of rVD and specific nucleotides. HD and nn monomers have strong hydrogen binding with nucleotides while nG and nI have weak binding. It determines possible binding of DNA-learning domain to the sites differing from target on several nucleotides. (2) There is also the possibility of interaction between nG and A due to degeneracy of a code. (3) Dimerization of FokI domains of two nucleases with the identical DNA-binding domains is possible. This issue was resolved in a series of experiments that involved receiving of TALENs which contain the FokI domains working as obligate heterodimers.

(4) Possible unintentional off-target effects can result from the fact that the size of spacer DNA between the sites of nucleases recognition is not fixed. This property results in a possible introduction of two-chained gaps during binding of nucleases with the non-target sites located at the distance sufficient for dimerization of FokI domains (Mussolino et al. 2011; Gaj et al. 2013). (5) Finally, non-target effects can result from similarity of the sequences between species of insects which haven't been determined using bioinformatics methods and/or single nucleotide polymorphisms (SNPs) (Perkin et al. 2016).

A majority of non-target effects are mitigated by careful planning that involves choosing guides with excellent on-target estimates, applying modified endonucleases with two guides, and also to enter changes into original CRISPR methods, that will allow designing specific gene cartridges specific to certain insect species (Graham and Root 2015; Perkin et al. 2016). In addition, an important experimental feature is a careful selection of sites for specific introduction of a two-chained gap. Selection of the required sites needs to avoid sites of repetitive sequences, as well as sites having a high degree of homology with other regions of the genome (Perkin et al. 2016).

Conclusion

In summary, both transgenic and RNAi-based insect-resistant crops and their products have no fundamental risks to human health. The potential risks as outlined in this chapter associated with genetic modification, are typical for most of the novel GM cultivars. As research studies have always shown, the rare deviations from the norm and from "expected" phenotype could be easily visualized and discarded at the earliest testing stages of all types of IR GM crops.

Presently, the GMOs (including insect-resistant crops) risk assessment system on human health and/or environment is strictly based on sound theoretical and experimental data. An ideal system of IR GM crop risk assessment includes the several stages and begins with the information analysis of organism's biological characteristics, which has been used for the generation of all types of IR GM cultivars (including the specifics of host organism, recipient organism, donor organism, parental organism). The obtained information is a basis for the modified organism's comparison with existing varieties and allows to determine the initial risk level. The next steps involve the analysis of the genetic modification protocol in detail that include the transgenic and/or RNAi/genome edited organism genetic engineering method; the transgenic construct features and vector source; an accurate description of recipient DNA fragment integrated into the genome; and insertion stability definition. The data obtained by analysis allows tracking exactly which genes are transferred to the original organism and determining the presence or absence of genes, causing the increased risk, such as antibiotic resistance marker genes, and the genes for the anti-nutritional substances synthesis. Such rigorous analysis reduces the level of scientific uncertainty related with potential unintended modification effects. For example, the inserted region and transgenes stability information are used to assess the likelihood of possible gene activity changes in the recipient-organism and the modified pleiotropic effects. Information on regulatory elements of transferred molecular structure permits judging of the expected expression level of target gene and its tissue specificity.

The next stage of risk assessment is to provide the biological characteristics of modified organism. The detailed description of the IR GM cultivars genotype and phenotype is conducted at this stage. It focuses on the features and characteristics that appear and/or disappear compared to the original wild-type organism. In addition, the genetic stability and the transgene expression level (including marker) in the IR GM plant genome data are considered. Information of the synthesis of transgene-encoded products and tissue specificity of this synthesis is taken into account. The obtained information, in a certain degree of probability, allows comparing the risk of the IR GM plant consumption with the existing risks of "traditional" organisms' (analogs) consumption. For example, changes in the toxicity potential and/or the allergenicity compared to the original wild-type analog, and allergens in the edible and inedible GM plants parts could be detected based on the toxins level of the transgene-encoded product and anti-nutritional substances. At the potential detection of toxic and/or allergenic properties manifestations, the detailed toxicological and allergological examination of the obtained transgene is carried out using methodological approach recommended by FAO/WHO.

In the final stage, the overall risk assessment caused by IR GM plants is conducted based on the likelihood of risk emergence and the identified adverse consequences, and if such adverse effects actually take place. The recommendations as to whether the risks are acceptable or controlled, including the strategies for identification of risk regulation are submitted for accurate reports. Implementation of the proposed approach(es) to the potential risks assessment of insect-resistant GM plants can eliminate or at least minimize any potential adverse effects of GMOs on human health and/or environment.

References

Abdurakhmonov, I.Y. 2016. RNA interference—a hallmark of cellular function and gene manipulation. pp. 1–18. *In*: Abdurakhmonov, I.Y. (ed.). RNA Interference. In-Tech.

Abdurakhmonov, I.Y., M.S. Ayubov, K.A. Ubaydulleava et al. 2016. RNA interference for functional genomics and improvement of cotton (*Gossypium* spp.). Front Plant Sci. 7: 202.

Bachman, P.M., R. Bolognesi, W.J. Moar et al. 2013. Characterization of the spectrum of insecticidal activity of a double-stranded RNA with targeted activity against western corn rootworm (*Diabrotica virgifera virgifera* LeConte). Transgenic Res. 22: 1207–1222.

Baktavachalam, G.B., B. Delaney, T.L. Fisher et al. 2015. Transgenic maize event TC1507: Global status of food, feed, and environmental safety. GM Crops Food. 6(2): 80–102.

Barrangou, R. and J.A. Doudna. 2016. Applications of CRISPR technologies in research and beyond. Nat Biotechnol. 34(9): 933–941.

Batista, R., B. Nunes, M. Carmo et al. 2005. Lack of detectable allergenicity of transgenic maize and soya samples. J Allergy Clin Immunol. 116(2): 403–410.

Baum, J.A., T. Bogaert, W. Clinton et al. 2007. Control of coleopteran insect pests through RNA interference. Nat Biotechnol. 25: 1322–1326.

Bergmans, H., C. Logie, K. Van Maanen et al. 2008. Identification of potentially hazardous human gene products in GMO risk assessment. Environ Biosafety Res. 7: 1–9.

Bernstein, J.A., I.L. Bernstein, L. Buchinni et al. 2003. Clinical and laboratory investigation of allergy to genetically modified foods. Environ Health Perspect. 111(8): 1114–21.

Betz, F.S., B.G. Hammonds and R.L. Fuchs. 2000. Safety and advantages of *Bacillus thuringiensis*-protected plants to control insect pests. Regul Toxicol Pharmacol. 32(2): 156–73.

Bravo, A., S.S. Gill and M. Soberon. 2007. Mode of action of *Bacillus thuringiensis* Cry and Cyt toxins and their potential for insect control. Toxicon. 49(4): 423–435.

Breyer, D., L. Kopertekh and D. Reheul. 2014. Alternatives to antibiotic resistance marker genes for *in vitro* selection of genetically modified plants—scientific developments, current use, operational access and biosafety considerations. Crit Rev Plant Sci. 33: 286–330.

Bucchini, L. and L.R. Goldman. 2002. Starlink corn: A risk analysis. Environ Health Perspect. 110(1): 5–13.

Casacuberta, J.M., Y. Devos, P. du Jardin et al. 2014. Biotechnological uses of RNAi in plants: risk assessment considerations. Trends Biotechnol. 33: 145–147.

Cellini, F., A. Chesson, I. Colquhon et al. 2004. Unintended effects and their detection in genetically modified crops. Food Chem Toxicol. 42: 1089–125.

Champer, J., A. Buchman and O.S. Akbari. 2016. Cheating evolution: Engineering gene drives to manipulate the fate of wild populations. Nat Rev Genet. 17: 146–159.

Chassy, B.M. 2002. Food safety evaluation of crops produced through biotechnology. J Amer Coll Nutr. 21(3): 166–173.

Clarke, J.D., D.C. Alexander, D.P. Ward et al. 2013. Assessment of genetically modified soybean in relation to natural variation in the soybean seed metabolome. Sci Rep. 3: 3082.

Creighton, T.E. 1993. Proteins: Structures and Molecular Properties. New York: W.H. Freeman and Company.

DeFrancesco, L. 2013. How safe does transgenic food need to be? Nat Biotechnol. 31: 794–802.

DeVries, J., P. Meier and W. Wackernagel. 2001. The natural transformation of the soil bacteria Pseudomonas stutzeri and *Acinetobacter* sp. by transgenic plant DNA strictly depends on homologous sequences in the recipient cells. FEMS Microbiol Lett. 195: 211–215.

EFSA Panel on Genetically Modified Organisms (GMO). 2012a. Scientific opinion addressing the safety assessment of plants developed through cisgenesis and intragenesis. EFSA Journal. 10: 2561.

EFSA Panel on Genetically Modified Organisms (GMO). 2012b. Scientific opinion addressing the safety assessment of plants developed using zinc finger nucleases and other site-directed nucleases with similar function. EFSA Journal. 10: 2943.

EFSA. International scientific workshop 'Risk assessment considerations for RNAibased GM plants'. 4–5 June 2014. Brussels, Belgium.

Espinoza, C., R. Schlecter, D. Herrera et al. 2013. Cisgenesis and intragenesis: New tools for improving crops. Biol Res. 46: 323–331.

FAO. 2013. The state of food and agriculture. Rome.

FAO. 2015. The State of Food Insecurity in the World. Meeting the 2015 international hunger targets: taking stock of uneven progress. Rome.

FAO/WHO. 2000. Safety aspects of genetically modified foods of plant origin: Report of a Joint FAO/WHO Expert Consultation on Foods Derived from Biotechnology. Food and Agriculture Organization of the United Nations and World Health Organization. WHO, Geneva, Switzerland.

FAO/WHO. 2001. Allergenicity of genetically modified foods. Report of a Joint FAO/WHO Expert Consultation on Foods Derived from Biotechnology. Food and Agriculture Organization of the United Nations, Rome. 1–29.

FAO/WHO. 2001a. Safety assessments of foods derived from genetically modified microorganisms. Report of a Joint FAO/WHO Expert Consultation on Foods Derived from Biotechnology. Food and Agriculture Organization of the United Nations and World Health Organization. WHO, Geneva, Switzerland.

FAO/WHO. 2001b. Evaluation of allergenicity of genetically modified foods. Report of a Joint FAO/WHO Expert Consultation on Allergenicity of Foods Derived from Biotechnology. Food and Agriculture Organization of the United Nations and World Health Organization. FAO, Rome, Italy.

FAO/WHO. 2003. Joint FAO/WHO Food Standards Programme «Codex Alimentarius Commission». Appendix II «Draft guideline for the conduct of food safety assessment of foods derived from recombinant-DNA plants». Twenty-Sixth Session (ALINORM 03/34), FAO Headquarters, Rome, 30 June–7 July 2003. Rome. pp. 47–57.

FAO/WHO. 2015. Codex Alimentarius. Procedural Manual.

FDA. 1992. Statement of policy: Foods derived from new plant varieties. Food and Drug Administration Fed Reg. 57: 22984–23002.

Federici, B.A. and J.P. Siegel. 2008. Safety assessment of *Bacillus thuringiensis* and Bt crops used in insect control. *In*: Hammond, B.G. (ed.). Food Safety of Proteins in Agricultural Biotechnology. Boca Raton, FL: CRC Press.

Fontes, E.M.G., C.S.S. Pires, E.R. Sujii et al. 2002. The environmental effects of genetically modified crops resistant to insects. Neotrop Entomol. 31(4): 497–513.

Gaj, T., C.A. Gerbach and C.F. Barbas. 2013. ZFN, TALEN, and CRISPR/Cas-based methods for genome engineering. Trends Biotechnol. 31(7): 397–405.

Gatehouse, A.M.R., N. Ferry, M.G. Edwards et al. 2011. Insect-resistant biotech crops and their impacts on beneficial arthropods. Phil Trans R Soc B. 366: 1438–1452.

Goodman, R.E. and A.O. Tetteh. 2011. Suggested improvements for the allergenicity assessment of genetically modified plants used in foods. Curr Allergy Asthma Rep. 11: 317–324.

Goodman, R.E. 2013. Evaluation of endogenous allergens for the safety evaluation of genetically engineered food crops: Review of potential risks, test methods, examples and relevance. J Agric Food Chem. 61: 8317–8332.

Graham, D.B. and D.E. Root. 2015. Resources for the design of CRISPR gene editing experiments. Genome Biol. 16: 260.

Gray, A. 2012. Problem formulation in environmental risk assessment for genetically modified crops: a practitioner's approach. Coll Biosafety Rev. 6: 10–65.

Guo, Q.-Y., L. He, H. Zhu et al. 2015. Effects of 90-day feeding of transgenic maize BT799 on the reproductive system in male wistar rats. Int J Environ Res Public Health. 12: 15309–15320.

Hammond, B.G. and J.M. Jez. 2011. Impact of food processing on the safety assessment for proteins introduced into biotechnology-derived soybean and corn crops. Food Chem Toxicol. 49(4): 711–21.

Hartley, S., F. Gillund, L. Van Love et al. 2016. Essential features of responsible governance of agricultural biotechnology. PLoS Biol. 14(5): e1002453.

Heinemann, J.A., S.Z. Agapito-Tenfen and J.A. Carman. 2013. A comparative evaluation of the regulation of GM crops or products containing dsRNA and suggested improvements to risk assessments. Environ Int. 55: 43–55.

Herman, R.A. and W.D. Price. 2013. Unintended compositional changes in genetically modified (GM) crops: 20 years of research. J Agric Food Chem. 61: 11695–11701.

Ho, M.-W. 2014. Horizontal transfer of GM DNA: why is almost no one looking? Open letter to Kaare Nielsen in his capacity as a member of the European Food Safety Authority GMO panel. Microbial Ecol Health Dis. 25: 25919.

Hong, B., T.L. Fisher, T.S. Sult et al. 2014. Model-based tolerance intervals derived from cumulative historical composition data: Application for substantial equivalence assessment of a genetically modified crop. J Agric Food Chem. 62: 9916–9926.

IFIC. 2001. Understanding food allergy. IFIC Review. 1–7.

Ivashuta, S., Y. Zhang, B.E. Wiggins et al. 2015. Environmental RNAi in herbivorous insects. RNA. 21: 840–850.

Jackson, A.L. 2006. Widespread siRNA "off-target" transcript silencing mediated by seed region sequence complementarity. RNA. 12: 1179–1187.

James, C. 2007. Global Status of Commercialized Biotech. GM crops. ISAAA Brief. 37. ISAAA: Ithaca, NY.

James, C. 2015. 20th Anniversary (1996 to 2015) of the Global Commercialization of Biotech Crops and Biotech Crop Highlights in 2015. ISAAA Brief. 51. ISAAA: Ithaca, NY.

Keese, P. 2008. Risks from GMOs due to horizontal gene transfer. Environ Biosafety Res. 7: 123–149.

Khatodia, S., K. Bhatotia, N. Passricha et al. 2016. The CRISPR/Cas genome-editing tool: Application in improvement of crops. Front Plant Sci. 7: Article 506.

Kim, J.S. and C.O. Pabo. 1998. Getting a handhold on DNA: design of poly-zinc finger proteins with femtomolar dissociation constants. Proc Natl Acad Sci USA. 95: 2812–2817.

Kleter, G.A., A.C.M. Peijneneberg and H.J.M. Aaarts. 2005. Health considerations regarding horizontal transfer of microbial transgenes present in genetically modified crops. J Biomed Biotechnol. 2005(4): 326–52.

Koch, M.S., J.M. Ward, S.L. Levine et al. 2015. The food and environmental safety of Bt crops. Front Plant Sci. 8: 283.

Kohl, C., G. Framton, J. Sweet et al. 2015. Can systematic reviews inform GMo risk assessment and risk management? Front Bioeng Biotechnol. 3: 113.

Kok, E.J., J. Keijer and A.M.A. Van Hoef. 1998. mRNA fingerprinting of transgenic food crops. pp. 37–49. *In*: Horning, M. (ed.). Report of the Demonstration Programme on Food Safety Evaluation of Genetically Modified Food as a Basis for Market Introduction. The Hague: Ministry of Economic Affairs.

Kuiper, H.A., G.A. Kleter, H.P. Noteborn et al. 2001. Assessment of the food safety issues related to genetically modified foods. Plant J. 27: 503–528.

Kumar, S., A. Chandra and K.C. Pandey. 2008. *Bacillus thuringiensis* (Bt) transgenic crop: An environment friendly insect-pest management strategy. J Environ Biol. 29(5): 641–653.

Ladics, G.S., A. Bartholomeus, P. Bregitzer et al. 2015. Genetic basis and detection of unintended effects in genetically modified crop plants. Transgenic Res. 24: 587–603.

Li, G., Y. Wang, B. Liu et al. 2014. Transgenic *Bacillus thuringiensis* (Bt) rice is safer to aquatic ecosystems than its non-transgenic counterpart. PLoS ONE. 9(8): e104270.

Li, Y. and J. Romeis. 2010. Bt maize expressing Cry3Bb1 does not harm the spider mite, *Tetranychus urticae*, or its ladybird beetle predator, *Stethorus punctillum*. Biol Control. 53: 337–344.

Lowder, L., A. Malzhan and Y. Qi. 2016. Rapid evolution of manifold CRISPR systems for plant genome editing. Front Plant Sci. 7: 1683.
Lundgren, J.G. and J.J. Duan. 2013. RNAi-based insecticidal crops: Potential effects on nontarget species. BioSci. 63: 657–665.
Luo, M., B. Gilbert and M. Ayliffe. 2016. Applications of CRISPR/Cas9 technology for targeted mutagenesis, gene replacement and stacking of genes in higher plants. Plant Cell Rep. 35(7): 1439–1450.
Mao, Y.-B., X. Tao, X. Xue et al. 2011. Cotton plants expressing CYP6AE14 double-stranded RNA show enhanced resistance to bollworms. Transgenic Res. 20: 665–673.
Maryanski, J.H. 1995. Food and drug administration policy for foods developed by biotechnology. pp. 12–22. *In*: Engel, K.-H., G.R. Takeoka and R. Teranishi (eds.). Genetically Modified Foods: Safety Issues. Washington: ACS Symposium Series No. 605. American Chemical Society.
Mei, Y., Y. Wang, H. Chen et al. 2016. Recent progress in CRISPR/Cas9 technology. J Genet Genom. 43(2): 63–75.
Mezzomo, B.P., A.L. Miranda-Vellela, L.C. Barbosa et al. 2016. Hematotoxicity and genotoxicity evaluations in Swiss mice intraperitoneally exposed to *Bacillus thuringiensis* (var kurstaki) spore crystals genetically modified to express individually Cry1Aa, Cry1Ab, Cry1Ac, or Cry2Aa. Environ Toxicol. 31(8): 970–8.
Moens, W. and J.M. Collard. 2002. GM Plants containing antibiotic resistance genes. Belgian Biosafety Server.
Mussolino, C., R. Morbitzer, F. Lugge et al. 2011. A novel TALE nuclease scaffold enables high genome editing activity in combination with low toxicity. Nucleic Acids Res. 39: 9283–9293.
Nakayachi, K. 2013. The unintended effects of risk-refuting information on anxiety. Risk Anal. 33(1): 80–91.
Nielsen, K.M., A.m. Bones, K. Smalla et al. 1998. Horizontal gene transfer from transgenic plants to terrestrial bacteria—a rare event? FEMS Microbial Rev. 22: 79–103.
Nielsen, K.M., T. Bonne and J.P. Townsend. 2014. Detecting rare gene transfer events in bacterial populations. Front Microbiol. 4: 41.
Nordlee, J.A., S.A. Taylor, J.P. Townsend et al. 1996. Identification of a brazil nut allergen in transgenic soybeans. N Engl J Med. 334: 688–692.
Noteborn, H.P.J.M., A. Lommen and J.M. Weseman. 1998. Chemical fingerprinting and *in vitro* toxicological profiling for the safety evaluation of transgenic food crops. pp. 51–79. *In*: Horning, M. (ed.). Report of the Demonstration Programme on Food Safety Evaluation of Genetically Modified Food as a Basis for Market Introduction. The Hague: Ministry of Economic Affairs.
OECD. 1993. Safety Evaluation of Foods Produced by Modern Biotechnology—Concepts and Principles, OECD, Paris.
OECD. 1997. Safety assessment of new foods: results of an OECD survey of serum banks for allergenicity testing, and use of databases. ICGB. (97)1: 1–34.
OECD. 2000. Report of the task force for the safety of novel foods and feeds. Provides an overview on scientific issues and current approaches to food safety assessment.
OECD. Consensus documents for the work on the safety of novel foods and feeds. These documents provide information on compositional considerations for new (GM) varieties of several food and feed crops.
Palma, L., D. Munoz, C. Berry et al. 2014. *Bacillus thuringiensis* toxins: An overview of their biocidal activity. Toxins. 6: 3296–3325.
Perkin, L.C., S.L. Adrianos and B. Oppert. 2016. Gene disruption technologies have the potential to transform stored product insect pest control. Insects. 7(3): 46.
Petrick, J.S., B. Brower-Toland, A.L. Jackson et al. 2013. Safety assessment of food and feed from biotechnology-derived crops employing RNA-mediated gene regulation to achieve desired traits: A scientific review. Regul Toxicol Pharmacol. 66: 167–176.

Pray, C.E., J. Huang, R. Hu et al. 2002. Five years of Bt cotton in China—the benefits continue. Plant J. 31: 423–430.

Puchta, H. 2016. Applying CRISPR/Cas for genome engineering in plants: the best is yet to come. Curr Opin Plant Biol. 36: 1–8.

Ramon, M., Y. Devos, A. Lanzoni et al. 2014. RNAi-based GM plants: food for thought for risk assessors. Plant Biotechnol J. 12: 1271–1273.

Randhawa, G.J., M. Singh and M. Grover. 2011. Bioinformatic analysis for allergenicity assessment of *Bacillus thuringiensis* Cry proteins expressed in insect-resistant food crops. Food Chem Toxicol. 49(2): 356–362.

Raybould, A. 2006. Problem formulation and hypothesis testing for environmental risk assessment of genetically modified crops. Environ Biosafety Res. 5: 119–126.

Ricroch, A.E., J.B. Berge and M. Kuntz. 2011. Evaluation of genetically engineered crops using transcriptomic, proteomic, and metabolomic profiling techniques. Plant Physiol. 155: 1752–61.

Rischer, H. and K.M. Oksman-Caldentey. 2006. Unintended effects in genetically modified crops: revealed by metabolomics? Trends Biotechnol. 24: 102–104.

Rizzi, A., N. Radaddi, C. Sorlini et al. 2012. The stability and degradation of dietary DNA in the gastrointestinal tract of mammals: implications for horizontal gene transfer and the biosafety of GMOs. Crit Rev Food Sci Nutr. 52(2): 142–161.

Roberts, A.F., Y. Devos, G.M.Y. Lemgo et al. 2015. Biosafety research for non-target organism risk assessment of RNAi-based GE plants. Front Plant Sci. 6: 958.

Ross, M.A. and C.A. Lembi. 1985. Applied weed science. Burgess Publishing Co., Minneapolis. p. 340.

Royal Society. 2002. Genetically Modified Plants for Food Use and Human Health—An Update. Policy Document 4/02. The Royal Society, London.

Rubio-Infante, N. and L. Moreno-Fierros. 2016. An overview of the safety and biological effects of *Bacillus thuringiensis* Cry toxins in mammals. J Appl Toxicol. 36(5): 630–48.

Ruzzini, A.C. and J. Clardy. 2016. Gene flow and molecular innovation in bacteria. Curr Biol. 26(18): R859–64.

Shan, Q., Y. Wang, J. Qi et al. 2013. Targeted genome modification of crop plants using a CRISPR-Cas system. Nat Biotechnol. 31(8): 686–688.

Siegel, J.P. 2001. The mammalian safety of *Bacillus thuringiensis*-based insecticides. J Invertebr Pathol. 77(1): 13–21.

Simó, C., C. Ibanez, A. Valdes et al. 2014. Metabolomics of genetically modified crops. Inter J Mol Sci. 15: 18941–18966.

Sjoblad, R.D., J.D. McClintock and R. Engler. 1992. Toxicological considerations for protein components of biological pesticide products. Regul Toxicol Pharmacol. 15: 3–9.

Smagghe, G. and L. Swevers. 2014. Editorial overview: pests and resistance—RNAi research in insects. Curr Opin Insect Sci. 6: 4–5.

Smalla, K., S. Borin, H. Heuler et al. 2000. Horizontal transfer of antibiotic resistance genes from transgenic plants to bacteria—are there new to fuel the debate? Proc. of the 6th International Simp. on the Biosafety of Genetically Modified Organisms. Saskatoon, Canada. 146–154.

Sprink, T., D. Ericksson, J. Schiemann et al. 2016. Regulatory hurdles for genome editing: process- vs. product-based approaches in different regulatory contexts. Plant Cell Rep. 35: 1493–1506.

Stewart, C.N., H.A. Richards and M.D. Halfhill. 2000. Transgenic plants and biosafety: science, misconceptions and public perception. Biotechniques. 29: 832–836, 838–843.

Tabashnik, B.E., M. Zhang, J.A. Fabrick et al. 2015. Dual mode of action of Bt proteins: protoxin efficacy against resistant insects. Scient Rep. 5: 15107.

Telem, R.S., S.H. Wani, N.B. Singh et al. 2013. Cisgenics—A sustainable approach for crop improvement. Curr Genom. 14: 468–476.

Townsend, J.A., D.A. Wright, R.J. Winfrey et al. 2009. High-frequency modification of plant genes using engineered zinc-finger nucleases. Nature. 459: 442–445.

Townsend, J.P., T. Bohn and K.M. Nielsen. 2012. Assessing the probability of detection of horizontal gene transfer events in bacterial populations. Front Microbiol Antimicrobials, Resistance and Chemotherapy. 3: 27.

US EPA. 2001. Bt Plant-Incorporated Protectants October 15, 2001 Biopesticides Registration Action Document.

Wang, E.H., Z. Yu, J. Hu et al. 2013. Effects of 90-day feeding of transgenic Bt rice TT51 on the reproductive system in male rats. Food Chem Toxicol. 62: 390–6.

Wang, L., X. Wang, X. Jin et al. 2015. Comparative proteomics of Bt-transgenic and non-transgenic cotton leaves. Proteome Sci. 13: 15.

Wang, M., D. Luan, L. Tu et al. 2015. Long noncoding RNAs and their proposed functions in fiber development of cotton (*Gossypium* spp.). New Phytol. 207: 1181–1197.

WHO. 1995. Application of principles of substantial equivalence to the safety of foods or food components from plants derived by modern biotechnology.

WHO. 2000. Safety aspects of genetically modified foods of plant origin.

Whyard, S., A.D. Singh and S. Wong. 2009. Ingested double-stranded RNAs can act as species-specific insecticides. Insect Biochem Mol Biol. 39: 824–832.

Wolt, J.D., P. Keese, A. Raybould et al. 2010. Problem formulation in the environmental risk assessment for genetically modified plants. Transgenic Res. 19: 425–436.

Woo, H.-J., S.B. Lee, Y. Qin et al. 2015. Generation and molecular characterization of marker-free Bt transgenic rice plants by selectable marker-less transformation. Plant Biotechnol Rep. 9(6): 351–360.

Xiong, J.-S., J. Ding and Y. Li. 2015. Genome-editing technologies and their potential application in horticultural crop breeding. Horticult Res. 2: 15019.

Yerger, L., G. Ambrozius and P. Bayer. 1990. Clinical immunology and allergology.

Zha, W., X. Peng, R. Chen et al. 2011. Knockdown of midgut genes by dsRNA-transgenic plant-mediated RNA interference in the hemipteran insect *Nilaparvata lugens*. PLos One. 6: e20504.

Zhang, F., M.L. Maeder, E. Unger-Wallace et al. 2010. High frequency targeted mutagenesis in Arabidopsis thaliana using zinc finger nucleases. Proc Natl Acad Sci USA. 107: 12028–12033.

Zotti, M.J. and G. Smagghe. 2015. RNAi technology for insect management and protection of beneficial insects from diseases: lessons, challenges and risk assessments. Neotrop Entomol. 44: 197–213.

CHAPTER 10

Towards a Holistic Integrated Pest Management

Lessons Learned from Plant-Insect Mechanisms in the Field

Xiomara Sinisterra-Hunter[1] and *Wayne B. Hunter*[2,*]

INTRODUCTION

Ag-biotechnology along with emerging gene-editing technologies for modifying gene structure, or gene expression, have revolutionized research fields across: plant and animal breeding, plant and animal pathology, crop improvement, virology, microbiology, entomology, biopharma, and biomedical research. RNA interference (RNAi) based technologies have been pivotal in the development of various research applications, in functional genomics; crop improvement; the management of arthropod pests and viral pathogens. RNAi technology involves experimental modulation of gene expression at the post-transcriptional level. More recently topically applied RNAi technologies have emerged, providing a temporary modification of transcript levels, without genetic modification of the host. Current discovery and application of a new gene editing technology, identified in bacteria: Clustered Regularly Interspaced Short Palindromic Repeats (CRISPR) and the CRISPR-associated protein 9 (Cas9), referred to as: (CRISPR/Cas9), has demonstrated exponential acceptance in the scientific community being applied across a wide variety of genetic applications. While RNAi technology can suppress transcripts of gene expression, the CRISPR/Cas9 system can perform precision insertions and deletions in the eukaryotic genome that are inheritable. These two technologies (RNAi and CRISPR/Cas9) provide unique advantages for agricultural uses including improving crops, and management of insect pests and pathogens. Ag-biotechnology provides

[1] Ag emerging Technologies Consultants, 679 NE Emerson St., Fort Pierce, FL, 34983.
[2] USDA, ARS, U.S. Horticultural Research Lab, 2001 S. Rock Road, Fort Pierce, FL 34945.
* Corresponding author: Wayne.hunter@ars.usda.gov

broad applications in crop improvement to address global food security. As the demand for food continues to increase, the importance of developing new and varied food crops should move to the forefront of the economic and social agendas of governments around the world. Social stability depends on economic growth. Agriculture provides about 1 in 5 jobs in developed nations. In developing nations, this number is significantly higher. Adoption of ag-biotechnology to expand the biodiversity of crops beyond food, will increase 'plant-made product' industries that may result in increased economic gain for farmers. A renewed focus to examine the genetic diversity from weeds and distantly related plants to currently grown crops, can provide solutions to a wide set of local and global problems in agriculture linked to drought, insect pests, opportunities in biofuel production, and human health challenges. One example of this growing market is the use of 'Plant-bioreactors' to produce 'Plant-made Biopharma products', like vaccines, or 'Plant-made industrial products', like ethanol or biofuels. Plant-made products for biomaterials (fibers, paper, bioplastics) have been shown to be feasible with lower production costs and relatively easy scale-up production to meet increasing and varied demands. Ultimately, acceptance and increased understanding, by governments and societies-at-large, of the benefits to be gained from using ag-biotechnology will determine their role in providing increased food, feeds, and fibers, which will increase social and economic stability.

Role of Chemistry in Crop Production

Prior to Modern Chemistry

The use of chemicals for pest and disease control has a long history dated to about 4500 ago, when Sumerians used Sulphur compounds to control insects and mites. Later, about 3200 years ago, it has been noted that Chinese farmers used arsenic compounds for combat body lice (Fishel 2013; Taylor et al. 2007; Ware and Whitacre 2004). Some of those compounds have endured the test of time and are still in use. For example, the Bordeaux Mixture, which is based on copper sulphate and lime and is still use to control various fungal diseases; while others like whale oil and arsenic are no longer used as pesticides (Fishel 2013; Smith and Secoy 1976; Ware and Whitacre 2004). More recently, up until the 1940s inorganic substances, such as sodium chlorate and sulphuric acid as well as nitrophenols, chlorophenols, creosote, naphthalene, ammonium sulphate, sodium arsenate and petroleum oils were used to combat fungal and insect pests, and as herbicides. But these products have several downsides including high rates of application, lack of selectivity and were often phytotoxic (Fishel 2013; Smith and Secoy 1976). Throughout most of the 1950s, consumers and most policy makers were not overly concerned about the potential health risks associated with using chemical pesticides because widespread chemical use resulted in food products which were much cheaper and seemingly without any collateral damage. There were no documented incidents of people dying or being seriously hurt by the proper use these new chemical formulations and pesticides (Reinhardt and Ganzel 2003a). The following decades brought the development of products much more targeted to the specific pests and diseases and less toxic for the human operators and the environment (Reinhardt and Ganzel 2003b; Ware and Whitacre 2004).

The 1970s and 1980s brought the introduction of the world's bestselling herbicide; glyphosate, the low use rate products like sulfonylurea and imidazolinone (imi) herbicides, as well as dinitroanilines and the aryloxyphenoxypropionate (fop) and cyclohexanediones (dim) families. These decades also usher the synthesis of a 3rd generation of pyrethroids, the introduction of avermectins, benzoylureas and the microbials, B_t (*Bacillus thuringiensis*) as a spray treatment for insect pest control. This period also saw the introduction of the triazole, morpholine, imidazole, pyrimidine and dicarboxamide families of fungicides. Since most of the agrochemicals introduced at that time had a single mode of action; their long-term use resulted in problems with pathogen/insect/weed resistance. New integrated pest management strategies were introduced to combat this negative effect (Ware and Whitacre 2004).

In the 1990s research activities concentrated on finding new members of existing chemical families with greater selectivity, lower persistence, and better environmental and toxicological profiles (Ware and Whitacre 2004). Emergence of gene-based targeting in 2000's to 2016 provides technology which can target a single arthropod species (Baum et al. 2007; Bachman et al. 2013; Fire et al. 1998; Whyard et al. 2009). Traditional chemical companies are now developing insecticides that are 'softer', more specific to taxonomic Orders or Families of arthropods, or other invertebrates (i.e., Miticides versus Insecticides), with shorter environmental persistence, and greater efficacy. Additionally, special care is devoted to ensuring safe use for food and feeds (Oberemok et al. 2015; Van Eenennaam and Young 2014) (Fig. 1). During this time gene-based targeting for pests and pathogens was developing as a safer management strategy which used short single-stranded DNA fragments as DNA insecticides (Kawai-Tayooka et al. 2004; Oberemok et al. 2013; Oberemok and Nyadar 2015; Oberemok and Skorokhod 2014), or RNA specific sequences (Fire et al. 1998; Hannon 2002; Timmons and Fire 1998), set the foundation for the field of "Highly Specific Pest Control" (Andrade and Hunter 2016; Baum et al. 2007; Bachman et al. 2013; Baum and Roberts 2014; Bélles 2010; Gu and Knipple

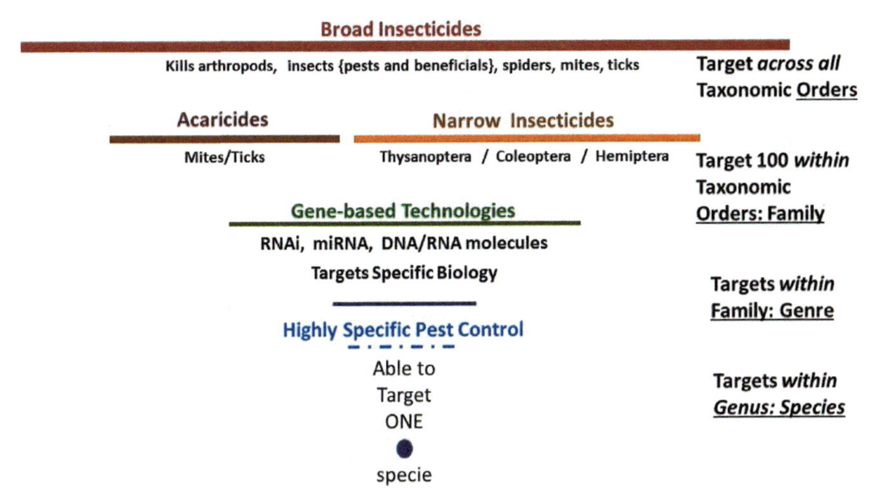

Figure 1. Technology continues to increase specificity of arthropod pest management. Continuum of improved chemical insecticides. Broad spectrum insecticides from the 1940's–1960's, followed by improved chemistries with more specific taxonomic effects 1970's–1990's and lower persistence. Emergence of gene-based targeting technology in 2000's to 2016 provides the opportunity to develop highly specific pest control products which can target single arthropod pest species.

2013; Huvenne and Smagghe 2010; Joga et al. 2016; Kanakala and Ghanim 2016; Thakur et al. 2016; Rodrigues and Figueira 2016; Wang et al. 2011; Whyard et al. 2009; Zotti and Smagghe 2015). These HiSPeC insecticides can appear to "think" before they act (Oberemok and Skorokhod 2014); meaning they can target one or two closely related insect pests within an entire ecosystem, whether it is wilderness, urban landscape, or cultivated field (Baum et al. 2007; Bachman et al. 2013; Gordon and Waterhouse 2007; Whyard et al. 2009) leaving beneficial insects and other organisms unharmed (Brutscher and Flenniken 2015; Hunter et al. 2010; Paldi et al. 2010; Garbian et al. 2012; Joga et al. 2016; Zotti and Smagghe 2015).

Nature is teaching humanity how to produce the best products for the management of pests and pathogens. Like the tale of "Goldilocks and the three bears": early developed chemistries that killed pathogens but were also toxic to humans, *"were too hot!"*. Chemistries that were safer, but used too often without appropriate rotation with other chemicals giving rise to insecticide resistance, *"were too cold!"*. The advances in genomics, and the increased understanding of natural gene expression systems, provides the knowledge and understanding that humanity has been searching for to develop methods that are *'just right!'*.

Future of Plant Breeding, Selection and Improvement

In the last 100 years, plant breeding and crop improvement has greatly advanced from the days when variety development was dependent upon crossing two parental lines, and then conducting years of selection from thousands of variants from field plantings. Methods to move desired genetic traits into plants one at a time has greatly shortened the time for improved crop development (Chaparro-Garcia et al. 2015; Lombardo et al. 2016; NAS 2016). Plant breeding has produced crops which have improved flavor and nutritional value, texture, and yields; crops have also been bred to have increased tolerance to drought, extreme temperatures, and to resist viruses, fungi and bacteria as well as insect pests. Development and adoption of genetic markers, as well as comparative genome analyses for plant quality traits have greatly shortened the time it takes to select improved crop varieties, but can still take 10–20 years to commercialize (Lobell et al. 2014; Sehgal et al. 2017; Van Eenennaam and Young 2014). Quality trait loci (QTLs) development and screening play a major role in modern breeding protocols directed to increasing production yields by providing a rapid tool to identify traits such as plant morphology, height, leaf size, roots mass; insect and pathogen resistance (Sehgal et al. 2017). These constant improvements in plant breeding, plant protection, and cultivation have led to today's world food production capacity which supports over 7 billion people (NAS 2016; Wise 2013a; Wise et al. 2013b).

Agricultural Biotechnology—Transgenic Technology

Transgenic technology refers to the alteration of the genetic machinery of living organisms (animals, plants or microorganisms). Mixing genes from different organisms is known as recombinant DNA technology and the resulting organism is said to be Genetically Modified (GM) or transgenic (Bawa and Anilakumar 2013). Transgenic technology has been at the forefront of developing insect pest-resistant plants. The first known commercial success story of a biotech crop was BT corn which was a corn variety transformed with an insecticidal protein from *Bacillus thuringiensis*. There are now commercially available Bt-cotton, Bt-maize, and Bt-rice. These biotech crops are resistant to a wide range of lepidopteran pests and exhibit well-documented extreme degree of antibiosis (Tabashnik et al. 2013). Biotech crops expressing insecticidal genes have help increase production in cotton, tobacco, potato, rice, strawberry, pea, Azuki bean, soybean, maize, and Poplar (Burand and Hunter 2013; Emani and Hunter 2013; NAS 2016).

Despite the commercial success, some concern exists among critics of the widespread adoption of biotech crops regarding the narrow focus (four major crops: soybean, maize, cotton, and canola) and the overreliance on two traits (insect resistance and herbicide tolerance). The last decade has seen a significant increase in biotech crop offerings including: sugar beet, alfalfa, squash, papaya, eggplant and poplar reaching a total of 10 biotech crops in 2015. To overcome pest resistant problems and to widen the target insect spectrum, Bt crops with extended transgene pyramiding (fusing non-toxic ricin B-chain molecules to the Bt transgene constructs) have been developed (Lombardo et al. 2016). Also, varieties with stacked traits (insect resistance and herbicide tolerance) continue to gain popularity. The adoption and area planted in biotech crops have steadily increased since their introduction in 1996, from few thousand hectares in the U.S.A. to over 179.1 million hectares in 28 countries, as recorded by The International Service for the Acquisition of Agri-Biotech Applications (ISAAA) in 2015 (Brookes and Barfoot 2016; James 2015). Controversies regarding the safety of biotech crops for human consumption and

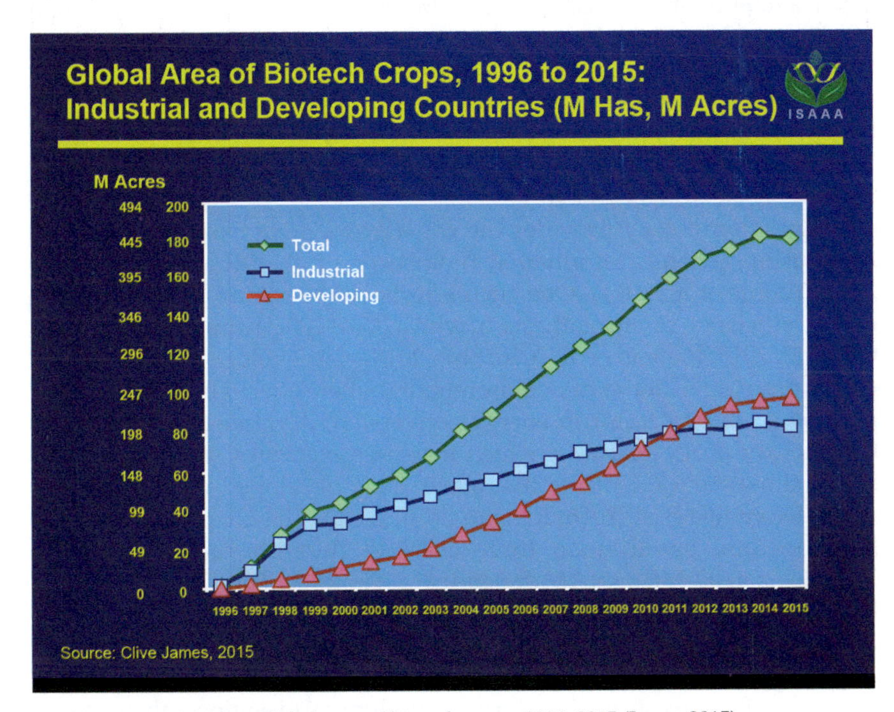

Figure 2. Global area of biotech crops, 1996–2015 (James 2015).

Norman Borlaug Quote:

"What we need is courage by the leaders of those countries where farmers still have no choice but to use older and less effective methods. The Green Revolution and now plant biotechnology are helping meet the growing demand for food production, while preserving our environment for future generations"
(ISAAA 2009).

environmental fate continue to hinder a much wider acceptance in the develop world specially in European Union country members which also prevents those in the developing world from reaping the social and economic benefits associated with products like golden rice which could eradicate malnutrition and blindness due to vitamin A deficiency in large swaths of Asia and Sub-Saharan Asia [The Golden Rice Project (http://www.goldenrice.org/)]. Biotech crops have shown their safety and utility to alleviate the effects of harsh environments by providing enhanced water and nitrogen efficiency, or drought tolerance (Brookes and Barfoot 2016; Koch et al. 2015; NAS 2016). It is important to emphasize that there is no evidence of any harmful effects caused by biotech crops. Among the chief complaints is the mistrust of the idea of putting genes from distantly related organisms into crop plants. However, gene transfer by microbes, viruses and bacteria, have been documented showing that this already occurs in nature. An international team of biologists discovered that fragments of the T-DNA of *Agrobacterium tumefasciens* had inserted into the genome of wild sweet potato (Kyndt et al. 2015). There is also a large body of publicly available data covering over 25 years of scientific research and analysis of biotech crops showing their safety, and economic benefits to farmers, consumers and the environment (Brookes and Barfoot 2016; Koch et al. 2015; Petrick 2013; Ricroch and Hénard-Damave 2016; Saurabh et al. 2014; NAS 2016).

Ag-biotechnology provides the capacity to 'adjust' or improve crop traits using genomic information (Lombardo et al. 2016; Saurabh et al. 2014; Younis et al. 2014). One such efforts is attempting to increase the energy efficiency, photosynthesis, of wheat (Gest 2002; Kromdijk et al. 2016; Ort et al. 2015; Rangan et al. 2016; Shih 2015). The evolutionary history of plants growing in hot or dry environments, where effective carbon supply may be a limiting factor in energy conversion, developed an improved C4 photosynthetic pathway, which increases efficient CO_2 capture. The study shows that C4 plants like maize, also developed greater growth rates (Kromdijk et al. 2016; Rangan et al. 2016). The C4 photosynthesis subtypes, are based upon three classical decarboxylation reactions include: NAD-ME (NAD-dependent malic enzyme); NADP-ME (NADP-dependent malic enzyme), and PEPCK (phosphoenolpyruvate carboxykinase) (Gowik and Westhoff 2011). A large portion of the world population depends on two main food crops, wheat and rice which are C3 plants. To increase yields, efforts are underway to engineer them to use the C4 pathway. Rangan et al. (2016), discovered that in wheat, currently classified as a C3 plant, there is a natural C4 photosynthetic pathway which operates in the developing wheat grain, and contributes about 30% to yield. However, the C4 pathway is absent in the leaves. Since the entire genetic C4 pathway is present and working in the grain the potential to reestablish the pathway throughout the leaves of wheat may be a reality. Increasing crop yields by adjusting existing gene pathways in wheat would not only produce more food, without increasing the need to cultivate more land, but would also demonstrate a new approach for improving yields in other C3 crops.

Agriculture is a labor-intensive endeavor, which provides one element for social stability, with 1 in 5 jobs being related to agriculture in the U.S.A. Efforts to repurpose endemic plants which are already adapted to a region may provide new crops with minimal investment, but may also increase jobs. Plants which were once thought of as weeds, or useless, may contain the genetic solutions for better crop varieties (Panepinto et al. 2015). Expanding the use of established crops as 'biofactories' for new 'plant-made products' like biofuels, is also becoming more popular. Currently use of low nicotine tobacco varieties as biofactories for biofuel are moving forward into commercialization (Ho et al. 2014; Sil and Jha 2014; Streatfield 2007; Panepinto et al. 2015; Poltonieri 2016). *"One acre of tobacco can yield up to 80 wet tons of biomass and all of the byproducts, including sugars, oils and proteins, can be used in products ranging from biofuel and animal feed to soil amendments (nutrients added to improve soil)"*. Tyton BioEnergy Systems.

Tobacco is a crop that is easy to grow, and can be grown on land not suitable for other crops, like soybean or maize, providing better use of available land, with low input costs, and significant yields and economic benefits (Grisan et al. 2016; Ho et al. 2014; Karlen et al. 2016; Sil and Jha 2014). Plant-made Biopharma products, like vaccines, which have been in development for many years are now making significant advances (Rosales-Mendoza 2015). A review by Sil and Jha (2014) describes bioactive molecules, produced in plant-made systems used for health and nutrition; sweetener, flavors, fragrances; insect repellents, and insecticides. One of the best-known examples is the treatment for breast, lung, and ovarian cancers, 'Taxol', which is produced from *Taxus brevifolia*, a conifer tree native to the Pacific Northwest of North America. Developing countries are finding new economic opportunities in 'Plant-made industrial products' like ethanol, or biofuels (Ho et al. 2014; Sil and Jha 2014; Vianna et al. 2011). The expansion and feasibility of plant-made products, with lower production costs, and relatively easy scale-up production, can provide new opportunities for growers in both developed and developing countries. Development, however depends upon governing agencies which need to develop effective regulatory guidelines and education programs to increase social acceptance (Grisan et al. 2016; Ho et al. 2014; Panepinto et al. 2015; Sil and Jha 2014).

Breakthroughs in agriculture and human medicine, which combine genomics and bioinformatics, have produced an -Omics generation of researchers who are rapidly discovering the benefits of understanding nature's own genetic processes, Life's biotechnology, for gene regulation, modification, and genetic editing, across a wide range of organisms from bacteria, plants, and animals (See reviews on: RNA interference (Dutta et al. 2015; Joga et al. 2016; Roberts et al. 2015; Lombardo et al. 2016; Li et al. 2015; Perkins et al. 2017; Whitten et al. 2016; Zhang et al. 2015; Zotti and Smagghe 2015); Zinc finger nucleases, ZNF; TAL effector nucleases, TALENS (Bortesi and Fischer 2015), and Clustered Regularly Interspaced Short Palindromic Repeats (CRISPR) and the CRISPR-associated protein 9 (Cas9), referred to as: (CRISPR/Cas9) system (Boettcher and McManus 2015; Dominguez et al. 2016; Gaj et al. 2013; Gupta and Shukla 2016; La Russa and Qi 2015; Liang et al. 2015; Wang et al. 2016; Wiedenheft et al. 2012; Wilson and Doudna 2013). Detailed descriptions of CRISPR system classification can be found in References (Koonin and Krupovic 2015; Makarova et al. 2015; Rath et al. 2015; Shmakov et al. 2015; Wang et al. 2016).

Plant Resistance to Insect Pests

Let's consider one strategy from nature that will surprise most people. Resistance to insects and other pests. Did you know that many plants produce their own pesticides? In a paper from 1990, environmental pioneer Bruce Ames wrote: *"About 99.9 percent of the chemicals humans ingest are made in nature. The amounts of synthetic pesticide residues in plant food are insignificant compared to the amount of natural pesticides produced by plants themselves. Of all dietary pesticides that humans eat, 99.99 percent are natural: they are chemicals produced by plants to defend themselves against fungi, insects, and other animal predators. We have estimated that on average Americans ingest roughly 5,000 to 10,000 different natural pesticides and their breakdown products. Americans eat about 1,500 mg of natural pesticides per person per day, which is about 10,000 times more than the 0.09 mg they consume of synthetic pesticide residues"* (Ames et al. 1990).

Examples of these natural pesticides are: *psoralen* which is produced by figs, parsley and celery and is toxic to insects and fish. *Solanine found in* potatoes, tomatoes, apples and okra has protective effects against fungi and blight. Cassava produces *cynanide* which protects the root from being eaten by insects and animals, but causes mortality in humans unless prepared properly. Borrowing from nature's playbook, breeders have sometimes tried to increase the amounts of pesticide produced naturally in order to make the plant more resistant to harm. Resilient plants provide the farmer a bigger harvest. Other times, breeders decrease the amount of toxins to make plants safer for human and animal consumption. However, this also makes the plant more susceptible to arthropods and animal herbivory. Classical breeding efforts are constantly being improved by molecular biological techniques. Bio-techniques are built upon the understanding of the genetic process in natural mechanisms, used by plants, insects, or microbes. Researchers then can adapt this information to produce DNA markers, genetic linkage maps, and molecular detection assays which enable taking tissue samples from a single seed without killing it, to determine if the desired traits are present. Using a three-pronged approach of antixenosis–antibiosis–tolerance, quantitative trait loci mapping (QTLs) related to both antixenotic and antibiotic resistance have been mapped in about 10 crop genera covering resistance to 21 arthropod species from the orders of Coleoptera, Hemiptera, and Lepidoptera (Sehgal et al. 2017; Smith and Clement 2012; Yencho et al. 2000). Some examples are: aphid resistance in wheat (Castro et al. 2005), midge resistance in sorghum (Tao et al. 2003), small brown plant hopper resistance in rice, corn earworm resistance in soybean (Rector et al. 2000), aphid and whitefly resistance in melon (Boissot et al. 2010), and aphid resistance in apple (Stoeckli et al. 2008).

Current and Future Crop Production

RNA Interference Applications for Crop Improvement

Multiple technologies for modifying gene structure, or gene expression, have revolutionized research fields across: plant and animal breeding, pathology, crop improvement, virology, microbiology, entomology, biopharma, and biomedical research, within the past 15 years (Boettcher and McManus 2015; Wang et al. 2016). RNA interference (RNAi) based technologies have dominated various research applications in functional genomics; development of improved crops (Lombardo et al. 2016; Saurabh et al. 2014; Younis et al. 2014); the management of arthropod pests (Galay et al. 2016; Joga et al. 2016; Kanakala and Ghanim 2016; Rodrigues and Figueira 2016; Thakur et al. 2016; Whyard et al. 2015; Xu et al. 2015; Zhang et al. 2015; Zotti and Smagghe 2015) and viral pathogens (Gonsalves et al. 2010; Jahan et al. 2011; Khan et al. 2013; Niu et al. 2014; Wang et al. 2006).

RNAi technology involves modulation of gene expression at the post-transcriptional level (Baulcombe 1996; Fire et al. 1998; Hannon 2002; Molnar et al. 2011). RNAi based technology provides a great opportunity to increase crop productivity across all organic, green, and traditional agricultural industries. Wider adoption of ag-biotech crops which are built upon natural biological systems, would greatly reduce the amounts of insecticides currently used (Koch and Kogel 2014; Koch et al. 2015; Lombardo et al. 2016; Prado et al. 2014; NAS 2016; Saurabh et al. 2014). Some of the first proof-of-concepts for plant-RNAi-mediated pest control were in genetically improved plants to nematodes (Yadav et al. 2006); Coleoptera, western corn rootworm, *Diabrotica virgifera virgifera* LeConte, maize (Baum et al. 2007), and Lepidoptera (*Helicoverpa armigera*) cotton, (Mao et al. 2007) pests. The technology was heralded as a big step forward in highly specific pest control (Baum et al. 2007; Bélles 2010; Borgio 2010; Huvenne and Smagghe 2010; Price and Gatehouse 2008; Turner et al. 2006). In crop plants, RNAi has traditionally been applied as a genetic insert for stable expression, producing a hairpin structure in genetically engineered plants (Gonsalves et al. 2010; Lombardo et al. 2016; Prado et al. 2014; Saurabh et al. 2014; Zha et al. 2011). A review by Saurabh et al. (2014) lists benefits from RNAi-mediated crop improvements, and applications which have been used for increasing crop resistance to viral pathogens, insect pests, to reduce toxins, or modify a plant trait. Success of RNAi crops which are engineered to express the RNAi trigger, such as transgenic Papaya are able to suppress papaya ringspot virus replication, enabling the plant to survive and produce fruit (Gonsalves et al. 2010; Saurabh et al. 2014). Zhang et al. (2015) also proposed restricting the RNAi event to the chloroplasts of the leaves, to target chewing pests, like the Colorado potato beetle (*Leptinotarsa decemlineata*). This approach would also enable targeting root pests, as specific plastids in the roots could express a dsRNA which could target nematodes (Dutta et al. 2015). The development and commercialization of genetically engineered crops often takes 10 or more years and millions of dollars (Baum and Roberts 2014; Palli 2014; Saurabh et al. 2014). An alternative option which is being investigated is the use of topically applied RNAi (Andrade and Hunter 2016; Baum and Roberts 2014; Hunter et al. 2012; Li et al. 2015; Palli 2014) to target pests and pathogens. Thus for some crops topical applied RNAi may provide a short term solution (Baum et al. 2007; Baum and Roberts 2014). Topically applied "host delivered RNAi" results by treating plant foliage, as a topically spray, or the root zone of plants, which absorb moving it systemically throughout the plant (Andrade and Hunter 2016; Hunter et al. 2012; Ivashuta et al. 2015; Li et al. 2015). RNAi provides several advantages. Topical RNAi only produces a temporary modification of the corresponding transcript level in the target organism, without genome modification of the treated plant, or host (Bachman et al. 2013; Baum and Roberts 2014; Ivashuta et al. 2015; Palli 2014; Roberts et al. 2015). The naturally high specificity of the RNAi mechanism is due to the accuracy needed for the cell's survival when processing nucleic acids whether for regulatory or defense (Elbashir et al. 2001; Ivashuta et al. 2015; Saleh et al. 2006). Therefore target specificity is built into RNAi technology when properly selecting a known gene sequence in the insect pest, plant, or pathogen (Baum et al. 2007; Fire et al. 1998; Whyard et al. 2009). Thus RNAi provides for the first time in human history of pest control, a management tool for highly specific pest suppression which targets only the corresponding RNA sequence in one, or a few closely related species (Baum et al. 2007; Bélles 2010; Bolognesi et al. 2012; Galay et al. 2016; Gordon and Waterhouse 2007; Gu and Knipple 2013; Huvenne and Smagghe 2010; Whyard et al. 2009) (see reviews: Bachman et al. 2013; Baum and Roberts 2014; Dutta et al. 2015; Joga et al. 2016; Saurabh et al. 2014; Thakur et al. 2016). Current use of RNAi provide the lowest risk to other organisms (Koch and Kogel 2014; Petrick et al. 2013; Roberts et al. 2015; Zhang et al. 2015). The high specificity of targeting, whether a viral pathogen, or a single insect pest, continues to be demonstrated (Baum et al. 2007; Bachman et al. 2013; Dutta et al. 2015; Hunter et al. 2010; Whyard et al. 2009). For example, the first outdoor trial with an RNAi product was focused on improving the health of honey bees, by suppressing the replication of a viral pathogen (Hunter et al. 2010). The dsRNA product (Remebee™-IAPV), targeted the Israeli acute paralysis virus which had been shown to cause bee decline (Maori et al. 2009). The delivery of the dsRNA was easily incorporated into the standard beekeeper protocols, where sucrose solutions are provided to bee hives during the winter months when forage is low, and bee stress is high. Following studies also demonstrated that the endoparasite, *Nosema*, and the ectoparasite, *Varroa*, could be reduced by feeding honey bees sucrose plus dsRNA targeting each parasite (Garbian et al. 2012; Paldi et al. 2010). Thus RNAi technology provides a wide range of applications from genetic engineering in plants (Abdurakhmonov et al. 2016; Lombardo et al. 2016; Saurabh et al. 2014), to the use of topical RNAi applications, soil treatments, or baits, where a short-term treatment is required, or where other technologies cannot be easily applied (Baum and Roberts 2014; Housden and Perrimon 2016; Lombardo et al. 2016; Unniyampurath et al. 2016; Wang et al. 2016). However, while topically applied RNAi provides an environmentally sound approach in pest and pathogen management, the characteristics which make it safer, such as complete degradation by UV sunlight (Li et al. 2015), and rapid degradation upon ingestion by non-target organism and microbes, within hours to days (Dubelman et al. 2014; Petrick et al. 2013); these properties also make effective RNAi treatments difficult to develop (Andrade and Hunter 2016; Baum and Roberts 2015; Bolognesi et al. 2012; Huvenne and Smagghe 2010; Joga et al. 2016; Li et al. 2015; Mao and Zeng 2012; Miller et al. 2012; Pridgeon et al. 2008; Scott et al. 2013; Terenius et al. 2011; Whitten et al. 2016).

Gene-editing Technologies CRISPR/Cas9, Applications for Crop Improvement

Current discovery and application of a new gene editing technology, identified in bacteria: Clustered Regularly Interspaced Short Palindromic Repeats (CRISPR) and the CRISPR-associated protein 9 (Cas9), referred to as: (CRISPR/Cas9) constitutes an adaptive immune system in many proteobacteria and archaea (Bhaya et al. 2011; Doudna and Charpentier 2014). CRISPR/Cas9 has demonstrated exponential acceptance in the scientific community being applied across a wide variety of organisms and genetic applications (Boettcher and McManus 2015; Char et al. 2017; Housden and Perrimon 2016; Perkins et al. 2017; Unniyampurath et al. 2016; Wang et al. 2016). While RNAi technology can suppress gene expression through transcript degradation; CRISPR/Cas9 can perform precision insertions and deletions in the eukaryotic genome that becomes a heritable trait (Bortesi and Fischer 2015; Boettcher and McManus 2015; Housden and Perrimon 2016; Wang et al. 2016). Both technologies are widely used as gene expression modifying research tools for various scientific discovery purposes. In general, both RNAi and CRISPR/Cas9 have applications which can be classified into four major domains: (i) research tool applications, (ii) crop improvement, (iii) agricultural animals, (iv) biomed/clinical applications. Only the CRISPR/Cas9 which targets the DNA, can introduce or correct mutations into genes. RNAi cannot be used for such precision genome engineering applications because it functions primarily by modulating protein production as a transcript degradation process (Boettcher and McManus 2015; Housden and Perrimon 2016; Unniyampurath et al. 2016). However, RNAi has a proven track record of success when used in transgenic plants to suppress viral pathogens of crops, or to alter gene expression to modify plant traits (Abdurakhmonov et al. 2016; Baum and Roberts 2014; Dutta et al. 2015; Saurabh et al. 2014). With the first transgenic plant incorporating an RNAi component in maize to the corn rootworm (Lombardo et al. 2016), time will tell if RNAi continues to provide significant benefits in crop protection.

As an emerging technology for gene modifications for crop protection and improvement, CRISPR/Cas9 has also demonstrated success in modifying plant gene expression (Mao et al. 2013; Puchta and Fauser 2013; Xie and Yang 2013). Most recent reports have been on the use of CRISPR/Cas9 for the non-transgenic method of gene knockout, to reduce or remove toxic proteins, or proteins linked to food spoilage (Bortesi and Fischer 2015; Waltz 2016). Expanding applications of CRISPR/Cas9 are the targeted insertion of transgenes, to produce biofactories of plants, plant cells, or algae, in the fields of metabolic engineering and molecular farming for the production of specific metabolites or proteins (Bortesi and Fischer 2015; Karlen et al. 2016; Jin et al. 2015; Rosales-Mendoza 2015; Sil and Jha 2014).

While the idea of introducing genetic modifications for the control of insect vectored pathogens, or insect pests is not new, the advances in gene-editing and modification make it more plausible than 20 years ago. Insect pest management by CRISPR/Cas9 is slowly developing as the requirements for success are more daunting than those for RNAi. CRISPR/Cas9 design depends upon knowing the gene structure within the target organism so having good genomic data greatly aids research efforts for proper construction of the CRISPR/Cas9 system (Doudna and Charpentier 2014; Wang et al. 2016). Another short-coming is that most insects being studied still have not yet developed the capacity for embryonic transformation (Bortesi and Fischer 2015; Boettcher and McManus 2015; Gregory et al. 2016). Thus, CRISPR/Cas9 has proven difficult in many insect species, especially within the Hemiptera, as attempts for embryonic transformation of eggs requires processing them within 10 minutes of being oviposited. Some of these difficulties may be solved by using insect-infecting virus delivery of the CRISPR/Cas9 system, or some other innovative approach. When CRISPR/Cas9 can be used in an insect system, such as with *Anopheles stephensi* mosquito, the Cas9-mediated gene drive research looks promising. So while the CRISPR/Cas9 system is more difficult to use than RNAi, progress in some insect systems are steadily moving from lab bench towards field trails, though some researchers suggest that the technology alone may not be sufficient for malaria mosquito eradication (Gantz et al. 2015). Complex biological systems will continue to require a shift in how researchers and plant breeders approach pest management and crop improvement issues (Bortesi and Fischer 2015; Boettcher and McManus 2015; Brookes and Barfoot 2016; Chaparro-Garcia et al. 2015; Gregory et al. 2016; Joga et al. 2016). The ability for plant breeding and pest management to become more precise will continue to depend upon the use of the advances made in genomics and biotechnology. Nature has already provided the tools, and genetic components, such as insect vectors for delivery, viral pathogens for expression, and their host genomes upon which to build the desired outcome. By increasing the understanding of all the components within each biological system, it becomes easier to develop improved crop varieties with resistance to viral pathogens or insect pests (Ali et al. 2015a,b; Baltes et al. 2015; Belhaj et al. 2013; Chaparro-Garcia et al. 2015; Hanley-Bowdoin et al. 2013).

Each biotechnology, provides some unique advantage for different applications, and they can actually be used to augment each other to resolve difficult biological questions, suggesting that the future will need RNAi and CRISPR/Cas9, as well as other available gene-editing /modifying biotechnology methods to address the growing needs of an increasing population (Boettcher and McManus 2015; Housden and Perrimon 2016; Koch et al. 2015; NAS 2016; Prado et al. 2014; Saurabh et al. 2014; Unniyampurath et al. 2016; Wang et al. 2016).

Targeting Viral Pathogens

CRISPR/Cas9 targeting of viruses that infect humans and mammals, in principle could be targeted with gene editing technologies to directly edit the genomic DNA of viruses, to eliminate the virus, like hepatitis B virus (Kennedy et al. 2015). Similarly, for RNA genome viruses, that integrate into their host, like HIV in humans (Zhang et al. 2015b), or Insect picorna-like viruses in honey bee (Maori et al. 2009), one might target and mutate the genetic elements integrated into the host genome. These applications are possible using CRISPR/Cas9, but are not possible using RNAi, which will only target and suppress viral RNA transcripts. While CRISPR/Cas9 offers enormous potential to directly inactivate DNA viral pathogens in humans and animals, RNAi currently stands as the most promising tool to target RNA viruses, especially in plants (Abdurakhmonov et al. 2016; Aragão and Faria 2009; Becker 2013; De Paula et al. 2015; Gonsalves et al. 2010; Hajeri et al. 2014; Kanakala and Ghanim 2016; Kumar et al. 2012; Lombardo 2016; Zhang et al. 2015). Eventually RNA targeting applications using CRISPR systems, may become more efficient (Price et al. 2015). CRISPR/Cas9 induces a change to the genome, however emerging CRISPR systems, such as a novel technique termed 'CRISPR interference' (CRISPRi), acts more like RNAi, and interferes with the transcription of RNA, providing a reversible and incomplete modification of gene expression (Bosher and Labouesse 2000; Gilbert et al. 2014; Konermann et al. 2015; Peters et al. 2015; Qi et al. 2013). CRISPRi can act on both coding and noncoding sequences (reviews: Dominguez et al. 2016; La Russa and Qi 2015). Studies demonstrate that although CRISPR systems may work, their use is an evolving challenge when editing or modifying RNA (Boettcher and McManus 2015; Doudna and Charpentier 2014; Liang et al. 2015; Swiech et al. 2015; Wang et al. 2016; Wilson and Doudna 2013).

Prospects for Genetic Engineering

The constant demand for food, feed, and fiber will only be met with the increased use of ag-biotechnology. Commercialization of plant-made products, such as biofuels, and biopharmaceutical products will become increasingly important as countries attempt to reduce cost of medications and energy needs.

Gene-editing techniques like RNAi and CRISPR/Cas9 are demonstrating safer approaches to the development of new crop varieties, increased economic gains for diversifying bio-industries. Ag-biotechnology presents an amazing opportunity for precision crop improvement, which will increase biodiversity within agricultural production systems. Plants will be treated more as 'patients' wherein the diagnosis and the treatment will be highly specific to the identified problem, whether pests, pathogen, or emerging social preference, such the size, color, or flavor of the crop. Improved crops will have increased tolerance to drought stress, flooding, herbivory, or pathogens. Increasing crop yields will be realized using many different approaches, such as altering the photosynthetic system to produce greater yields from C3 plants without having to increase cultivated land (Kromdijk et al. 2016; Ort et al. 2015). Concerns with food security, and nutrition will more easily be addressed for those who need it most. Adoption and acceptance of plant biotechnology has grown every year since their introduction, but with only a few improved food crops available. The future will see a tremendous increase in the variety and diversity of crop traits, varieties, and species, aimed at meeting the present and future needs in developing countries. International companies will also increase use of plant-made biotechnology systems which will rapidly increase opportunities for economic gains in countries which embrace these emerging markets. Plant biofactories will produce biofuels, biopharmaceuticals, and novel biomaterials (Grisen et al. 2016; Poltonieri 2016).

However, as with all new technologies, there will be some resistance until they are shown to be safe and economically feasible (Prado et al. 2014; Sil and Jha 2014). As the economic and health benefits for improved crops, continue to be shown as safe and economically feasible, societies and governments will need to reevaluate how stringently new crops are regulated. Before the world can reap the benefits of biotechnology, the plant products need to be evaluated on their safety, and not judged by the methods used to produce them. Otherwise, the potential benefits from these technologies may not live up to their full potential (James 2015; Ricroch and Hénard-Damave 2016).

References

Abdurakhmonov, I.Y., M.S. Ayubov, K.A. Ubaydullaeva et al. 2016. RNA interference for functional genomics and improvement of cotton (*Gossypium* sp.). Front Plant Sci. 7: 202.

Ali, Z., A. Abulfaraj, A. Idris et al. 2015a. CRISPR/Cas9-mediated viral interference in plants. Genome Biol. 16: 238.

Ali, Z., A. Abul-Faraj, L. Li et al. 2015b. Efficient virus-mediated genome editing in plants using the CRISPR/Cas9 system. Mol Plant. 8: 1288–1291.

Ames, B.N., M. Profet and L.S. Gold. 1990. Dietary pesticides (99.9% all natural). Proc Natl Acad Sci. 87: 7777–7781.

Andrade, E.C. and W.B. Hunter. 2016. RNA Interference—Natural gene-based technology for highly specific pest control (HiSPeC). *In*: Dr. Ibrokhim Y. Abdurakhmonov (ed.). RNA Interference. InTech.

Aragão, F.J.L. and J.C. Faria. 2009. First transgenic geminivirus-resistant plant in the field. Nat Biotechnol. 27: 1086–1088.

Bachman, P., R. Bolognesi, W.J. Moar et al. 2013. Characterization of the spectrum of insecticidal activity of a double-stranded RNA with targeted activity against western corn rootworm *Diabrotica virgifera virgifera* LeConte. Transgenic Res. 22: 1207–1222.

Baltes, N.J., A.W. Hummel, E. Konecna et al. 2015. Conferring resistance to geminiviruses with the CRISPR-Cas prokaryotic immune system. Nat Plants. 1: 15145.
Baulcombe, D.C. 1996. RNA as a target and an initiator of post-transcriptional gene silencing in transgenic plants. Plant Mol Biol. 32: 79–88.
Baum, J.A., T. Bogaert, W. Clinton et al. 2007. Control of coleopteran insect pests through RNA interference. Nat Biotechnol. 25: 1322–1326.
Baum, J.A. and J.K. Roberts. 2014. Progress towards RNAi-mediated insect pest management. Adv Insect Physiol. 47: 249–295.
Bawa, A.S. and K.R. Anilakumar. 2013. Genetically modified foods: safety, risks and public concerns—a review. J Food Sci Technol. 50: 1035–1046.
Becker, A. (ed.). 2013. Virus-Induced Gene Silencing: Methods and Protocols, Methods in Molecular Biology, Vol. 975, DOI 10.1007/978-1-62703-278-0_1. Springer, New York.
Belhaj, K., A. Chaparro-Garcia, S. Kamoun et al. 2013. Plant genome editing made easy: targeted mutagenesis in model and crop plants using the CRISPR/Cas system. Plant Meth. 9: 39.
Bélles, X. 2010. Beyond Drosophila: RNAi *in vivo* and functional genomics in insects. Annu Rev Entomol. 55: 111–128.
Bhaya, D., M. Davison and R. Barrangou. 2011. CRISPR-Cas systems in bacteria and archaea: versatile small RNAs for adaptive defense and regulation. Annu Rev Genet. 45: 273–297.
Boettcher, M. and M.T. McManu. 2015. Choosing the right tool for the job: RNAi, TALEN, or CRISPR. Mol Cell. 58: 575–585.
Boissot, N., S. Thomas, N. Sauvion et al. 2010. Mapping and validation of QTLs for resistance to aphids and whiteflies in melon. Theor Appl Genet. 121: 9–20.
Bolognesi, R., P. Ramaseshadri, J. Anderson et al. 2012. Characterizing the mechanism of action of double-stranded RNA activity against western corn rootworm (*Diabrotica virgifera virgifera* LeConte). PLoS ONE. 7: e47534.
Borgio, J.F. 2010. RNAi mediated gene knockdown in sucking and chewing insect pests. J Biopesti. 3: 386–393.
Bortesi, L. and R. Fischer. 2015. The CRISPR/Cas9 system for plant genome editing and beyond. Biotechnol Adv. 33: 41–52.
Bosher, J.M. and M. Labouesse. 2000. RNA interference: Genetic wand and genetic watchdog. Nat Cell Biol. 2: E31–E36.
Brookes, G. and P. Barfoot. 2016. Global income and production impacts of using GM crop technology 1996–2014. GM Crops & Food. 7: 38–77.
Brutscher, L.M. and M.L. Flenniken. 2015. RNAi and antiviral defense in the honey bee. J Immunol Res. 2015: 941897. doi:10.1155/2015/941897.
Burand, J.P. and W.B. Hunter. 2013. RNAi: Future in insect management. J Invertebr Pathol. 112 suppl: S68–74.
Castro, A.M., A. Vasicek, M. Manifiesto et al. 2005. Mapping antixenosis genes on chromosome 6A of wheat to greenbug and to a new biotype of Russian wheat aphid. Plant Breed. 124: 229–323.
Chaparro-Garcia, A., S. Kamoun and V. Nekrasov. 2015. Boosting plant immunity with CRISPR/Cas. Genome Biol. 16: 254.
Char, S.N., A.K. Neelakandan, H. Nahampun et al. 2017. An Agrobacterium-delivered CRISPR/Cas9 system for high-frequency targeted mutagenesis in maize. Plant Biotechnol J. 15: 257–268.
De Paula, N.T., J.C. de Faria and F.J. Aragão. 2015. Reduction of viral load in whitefly (Bemisia tabaci Gen.) feeding on RNAi-mediated bean golden mosaic virus resistant transgenic bean plants. Virus Res. 210: 245–247.
Dominguez, A.A., W.A. Lim and L.S. Qi. 2016. Beyond editing: repurposing CRISPR-Cas9 for precision genome regulation and interrogation. Nat Rev Mol Cell Biol. 17: 5–15.
Doudna, J.A. and E. Charpentier. 2014. Genome editing. The new frontier of genome engineering with CRISPR-Cas9. Science. 346: 1258096.

Dutta, T.K., P. Banakar and U. Rao. 2015. The status of RNAi-based transgenic research in plant nematology. Frontiers in Microbiology. 5: 1–7.

Elbashir, S.M., W. Lendeckel and T. Tuschl. 2001. RNA interference is mediated by 21- and 22-nucleotide RNAs. Genes Dev. 15: 188–200.

Emani, C. and W. Hunter. 2013. Insect Resistance. Chapter 9. *In*: Kole, C. (ed.). Genomics and Breeding for Climate-Resilient Crops, Vol. 2, Springer-Verlag Berlin Heidelberg.

Fire, A., S. Xu, M.K. Montgomery et al. 1998. Potent and specific genetic interference by double-stranded RNA in *Caenorhabditis elegans*. Nature. 391: 806–811.

Fishel, F.M. 2013. Pest Management and Pesticides: A historical Perspective. Agronomy Department, UF/IFAS Extension, P1219.

Gaj, T., C.A. Gersbach, C.F. Barbas III. 2013. ZFN, TALEN, and CRISPR/Cas-based methods for genome engineering. Trends Biotechnol. 31: 397–405.

Galay, R.L., R. Umemiya-Shirafuji, M. Mochizuki et al. 2016. RNA interference—A powerful functional analysis tool for studying tick biology and its control. *In*: Dr. Ibrokhim Y. Abdurakhmonov (ed.). RNA Interference. InTech.

Gantz, V.M., N. Jasinskieneb, O. Tatarenkovab et al. 2015. Highly efficient Cas9-mediated gene drive for population modification of the malaria vector mosquito *Anopheles stephensi*. Proc Natl Acad Sci USA. 112: E6736–6743.

Garbian, Y., E. Maori, H. Kalev et al. 2012. Bidirectional transfer of RNAi between honeybee and *Varroa destructor*: *Varroa* gene silencing reduces *Varroa* population. PLoS Pathog. 8: e1003035.

Gest, H. 2002. History of the word photosynthesis and evolution of its definition. Photosyn Res. 73: 7–10.

Gilbert, L.A., M.A. Horlbeck, B. Adamson et al. 2014. Genome-scale CRISPR-mediated control of gene repression and activation. Cell. 159: 647–661.

Gonsalves, D., S. Tripathi, J.B. Carr et al. 2010. Papaya Ringspot virus. Plant Health Instr doi: 10.1094/PHI-I-2010-1004-01.

Gordon, K.H. and P.M. Waterhouse. 2007. RNAi for insect-proof plants. Nat Biotechnol. 25: 1231–1232.

Gowik, U. and P. Westhoff. 2011. The path from C3 to C4 photosynthesis. Plant Physiol. 155: 56–63.

Gregory, M., L. Alphey, N.I. Morrison et al. 2016. Insect transformation with *piggyBac*: getting the number of injections just right. Insect Mol Biol. 25: 259–271.

Grisan, S., R. Polizzotto, P. Raiola et al. 2016. Alternative use of tobacco as a sustainable crop for seed oil, biofuel, and biomass. Agron Sustain Dev. 36: 55.

Gu, L. and D.C. Knipple. 2013. Recent advances in RNA interference research in insects: implications for future insect pest management strategies. Crop Prot. 45: 36–40.

Gupta, S.K. and P. Shukla. 2016. Gene editing for cell engineering: trends and applications. Critical Rev Biotechnol. 37: 672–684.

Hajeri, S., N. Killiny, C. El-Mohtar et al. 2014. Citrus tristeza virus-based RNAi in citrus plants induces gene silencing in *Diaphorina citri*, a phloem-sap sucking insect vector of citrus greening disease (Huanglongbing). J Biotechnol. 176: 42–49.

Hanley-Bowdoin, L., E.R. Bejarano, D. Robertson et al. 2013. Geminiviruses: masters at redirecting and reprogramming plant processes. Nat Rev Microbiol. 11: 777–88.

Hannon, G.J. 2002. RNA interference. Nature. 418: 244–251.

Ho, D.P., H.H. Ngo and W. Guo. 2014. A mini review on renewable sources for biofuel. Bioresour Technol. 169: 742–749.

Housden, B.E. and N. Perrimon. 2016. Comparing CRISPR and RNAi-based screening technologies. Nat Biotechnol. 34: 621–623.

Hunter, W., J. Ellis, D. VanEngelsdorp et al. 2010. Large-scale field application of RNAi technology reducing Israeli acute paralysis virus disease in honey bees (*Apis mellifera*, Hymenoptera: Apidae). PLoS Path. 6: e1001160.

Hunter, W.B., E. Glick, N. Paldi et al. 2012. Advances in RNA interference: dsRNA treatment in trees and grapevines for insect pest suppression. Southwest Entomol. 37: 85–87.

Huvenne, H. and G. Smagghe. 2010. Mechanisms of dsRNA uptake in insects and potential of RNAi for pest control: a review. J Insect Physiol. 56: 227–235.
Ivashuta, S., Y. Zhang, B.E. Wiggins et al. 2015. Environmental RNAi in herbivorous insects. RNA. 21: 840–50.
Jahan, S., S. Khaliq, B. Samreen et al. 2011. Effect of combined siRNA of HCV E2 gene and HCV receptors against HCV. Virol J. 8: 295.
James, C. 2015. Global status of commercialized biotech/GM crops 2015: ISAAA Brief No. 51. Ithaca (NY): ISAAA.
Ji, X., H. Zhang, Y. Zhang et al. 2015. Establishing a CRISPR-Cas-like immune system conferring DNA virus resistance in plants. Nat Plants. 1: 15144.
Joga, M.R., M.J. Zotti, G. Smagghe et al. 2016. RNAi efficiency, systemic properties, and novel delivery methods for pest insect control: What we know so far. Front Physiol. 7: 553.
Kanakala, S. and M. Ghanim. 2016. RNA interference in insect vectors for plant viruses. Viruses. 8: 329.
Karlen, S.D., C. Zhang, M.L. Peck et al. 2016. Monolignol ferulate conjugates are naturally incorporated into plant lignins. Sci Adv. 2: e1600393.
Kawai-Toyooka, H., C. Kuramoto, K. Orui et al. 2004. DNA interference: a simple and efficient gene-silencing system for high-throughput functional analysis in the fern *Adiantum*. Plant Cell Physiol. 45: 1648–1657.
Kennedy, E.M., A.V. Kornepati and B.R. Cullen. 2015. Targeting hepatitis B virus cccDNA using CRISPR/Cas9. Antivir Res. 123: 188–192.
Khan, A.M., M. Ashfaq, Z. Kiss et al. 2013. Use of recombinant Tobacco mosaic virus to achieve RNA interference in plants against the Citrus Mealybug, *Planococcus citri* (Hemiptera: Pseudococcidae). PLoS ONE. 8: e73657.
Koch, A. and K.H. Kogel. 2014. New wind in the sails: improving the agronomic value of crop plants through RNAi-mediated gene silencing. Plant Biotechnol. 12: 821–831.
Koch, M.S., J.M. Ward, S.L. Levine et al. 2015. The food and environmental safety of Bt crops. Front Plant Sci. 6: 283.
Konermann, S., M.D. Brigham, A.E. Trevino et al. 2015. Genome-scale transcriptional activation by an engineered CRISPR-Cas9 complex. Nature. 517: 583–588.
Koonin, E.V. and M. Krupovic. 2015. Evolution of adaptive immunity from transposable elements combined with innate immune systems. Nat Rev Genet. 16: 184–192.
Kromdijk, J., K. Głowacka, L. Leonelli et al. 2016. Improving photosynthesis and crop productivity by accelerating recovery from photoprotection. Science. 354: 857–861.
Kumar, P., S.S. Pandit and I.T. Baldwin. 2012. Tobacco rattle virus vector: A rapid and transient means of silencing *Manduca sexta* genes by plant mediated RNA interference. PLoS ONE. 7: e31347.
Kyndt, T., D. Quispe, H. Zhai et al. 2015. The genome of cultivated sweet potato contains A. *tumefaciens* T-DNAs with expressed genes: An example of a naturally transgenic food crop. Proc Natl Acad Sci USA. 112: 5844–5849.
La Russa, M.F. and L.S. Qi. 2015. The new state of the art: Cas9 for gene activation and repression. Mol Cell Biol. 35: 3800–3809.
Li, H., R. Guan, H. Guo et al. 2015. New insights into an RNAi approach for plant defence against piercing-sucking and stem-borer insect pests. Plant Cell Env. 38: 2277–2285.
Liang, X., J. Potter, S. Kumar et al. 2015. Rapid and highly efficient mammalian cell engineering via Cas9 protein transfection. J Biotechnol. 208: 44–53.
Lobell, D.B., M.J. Roberts, W. Schlender et al. 2014. Greater sensitivity to drought accompanies maize yield increase in the U.S. Midwest. Science. 344: 516–519.
Lombardo, L., G. Coppola and S. Zelasco. 2016. New technologies for insect-resistant and herbicide-tolerant plants. Trends Biotechnol. 34: 49–57.
Makarova, K.S., Y.I. Wolf, O.S. Alkhnbashi et al. 2015. An updated evolutionary classification of CRISPR-Cas systems. Nat Rev Microbiol. 13: 722–736.
Mao, J. and F. Zeng. 2012. Feeding-based RNA interference of a gap gene is lethal to the pea aphid, *Acyrthosiphon pisum*. PLoS ONE. 7: e48718.

Mao, Y., H. Zhang, N. Xu et al. 2013. Application of the CRISPR-Cas system for efficient genome engineering in plants. Mol Plant. 2013: 2008–2011.

Mao, Y.B., W.J. Cai, J.W. Wang et al. 2007. Silencing a cotton bollworm P450 monooxygenase gene by plant-mediated RNAi impairs larval tolerance of gossypol. Nat Biotechnol. 25: 1307–1313.

Maori, E., N. Paldi, S. Shafir et al. 2009. IAPV, a bee-affecting virus associated with colony collapse disorder can be silenced by dsRNA ingestion. Insect Mol Biol. 18: 55–60.

Miller, S.C., K. Miyata, S.J. Brown et al. 2012. Dissecting systemic RNA interference in the red flour beetle Tribolium castaneum: Parameters affecting the efficiency of RNAi. PLoS ONE. 7: e47431.

Molnar, A., C. Melnyk and D.C. Baulcombe. 2011. Silencing signals in plants: a long journey for small RNAs. Genome Biol. 12: 215.

Niu, Y., B. Shen, Y. Cui et al. 2014. Generation of gene-modified cynomolgus monkey via Cas9/RNA-mediated gene targeting in one-cell embryos. Cell. 156: 836–843.

Oberemok, V.V., P.M. Nyadar, O.S. Zaytsev et al. 2013. Pioneer evaluation of the possible side effects of the DNA insecticides on wheat (*Triticum aestivum* L.). Int J Biochem Biophys. 1: 57–63.

Oberemok, V.V. and O.A. Skorokhod. 2014. Single-stranded DNA fragments of insect-specific nuclear polyhedrosis virus act as selective DNA insecticides for gypsy moth control. Pesticide Biochem Physiol. 113: 1–7.

Oberemok, V.V. and P.M. Nyadar. 2015. Investigation of mode of action of DNA insecticides on the basis of LdMNPV IAP-3 Gene. Turkish J Biol. 39: 258–264.

Oberemok, V.V., A.S. Zaitsev, N.N. Levchenko et al. 2015. A brief review of most widely used modern insecticides and prospects for the creation of DNA insecticides. Entomol Rev. 95: 824.

Ort, D.R., S.S. Merchant, J. Alric et al. 2015. Redesigning photosynthesis to sustainably meet global food and bioenergy demand. Proc Natl Acad Sci USA. 112: 8529–8536.

Paldi, N., E. Glick, M. Oliva et al. 2010. Effective gene silencing of a microsporidian parasite associated with honey bee (*Apis mellifera*) colony declines. Appl Env Microbiol. 76: 5960–5964.

Palli, S.R. 2014. RNA interference in Colorado potato beetle: steps toward development of dsRNA as a commercial insecticide. Curr Opin Insect Sci. 6: 1–8.

Panepinto, D., F. Viggiano and G. Genon. 2015. Energy production from biomass and its relevance to urban planning and compatibility assessment: two applicative cases in Italy. Clean Techn Environ Policy. 17: 1429.

Perkin, L.C., S.L. Adrianos and B. Oppert. 2016. Gene disruption technologies have the potential to transform stored product insect pest control. Insect. 7 pii: E46.

Peters, J.M., M.R. Silvis, D. Zhao et al. 2015. Bacterial CRISPR: accomplishments and prospects. Curr Opin Microbiol. 27: 121–126.

Petrick, J.S., B. Brower-Toland, A.L. Jackson et al. 2013. Safety assessment of food and feed from biotechnology-derived crops employing RNA-mediated gene regulation to achieve desired traits: a scientific review. Regul Toxicol Pharmacol. 66: 167–176.

Poltonieri, P. 2016. Tobacco seed oil for biofuels. *In*: Poltronieri, P. and O.F. D'Urso (eds.). BioTransformation of Agricultural Waste and By-Products. The Food, Feed, Fibre, Fuel (4F) Economy, Elsevier.

Prado, J.R., G. Segers, T. Voelker et al. 2014. Genetically engineered crops: from idea to product. Annu Rev Plant Biol. 65: e790.

Price, A.A., T.R. Sampson, H.K. Ratner et al. 2015. Cas9-mediated targeting of viral RNA in eukaryotic cells. Proc Natl Acad Sci USA. 112: 6164–6169.

Price, D.R. and J.A. Gatehouse. 2008. RNAi-mediated crop protection against insects. Trends Biotechnol. 26: 393–400.

Pridgeon, J.W., L. Zhao, J.J. Becnel et al. 2008. Topically applied AaeIAP1 double-stranded RNA kills female adults of *Aedes aegypti*. J Med Entomol. 45: 414–420.

Puchta, H. and F. Fauser. 2013. Gene targeting in plants: 25 years later. Int J Dev Biol. 57: 629–37.
Qi, L.S., M.H. Larson, L.A. Gilbert et al. 2013. Repurposing CRISPR as an RNA-guided platform for sequence-specific control of gene expression. Cell. 152: 1173–1183.
Rangan, P., A. Furtado and R.J. Henry. 2016. New evidence for grain specific C4 photosynthesis in wheat. Sci Rep. 6: 31721.
Rath, D., L. Amlinger, A. Rath et al. 2015. The CRISPR-Cas immune system: biology, mechanisms and applications. Biochimie. 117: 119–128.
Rector, B.G., J.N. Allb, W.A. Parrottc et al. 2000. Quantitative trait loci for antibiosis resistance to corn earworm in soybean. Crop Sci. 40: 233–238.
Reinhardt, C. and B. Ganzel. 2003a. Wessels Living History Farm, York, Nebraska; Farming in the 1950s & 60s. The Ganzel Group.
Reinhardt, C. and B. Ganzel. 2003b. Wessels Living History Farm, York, Nebraska; Farming in the 1930s. The Ganzel Group.
Ricroch, A.E. and M.C. Hénard-Damave. 2016. Next biotech plants: new traits, crops, developers and technologies for addressing global challenges. Crit Rev Biotechnol. 36: 675–690.
Roberts, A.F., Y. Devos, G.N.Y. Lemgo et al. 2015. Biosafety research for non-target organism risk assessment of RNAi-based GE plants. Front Plant Sci. 6: 958.
Rodrigues, T.B. and A. Figueira. 2016. Management of insect pest by RNAi—A new tool for crop protection. *In*: Dr. Ibrokhim Y. Abdurakhmonov (ed.). RNA Interference. InTech.
Rosales-Mendoza, S. 2015. Current developments and future prospects for plant-made biopharmaceuticals against rabies. Mol Biotechnol. 57: 869–879.
Saleh, M.C., R.P. van Rij, A. Hekele et al. 2006. The endocytic pathway mediates cell entry of dsRNA to induce RNAi silencing. Nat Cell Biol. 8: 793–802.
Saurabh, S., A.S. Vidyarthi and D. Prasad. 2014. RNA interference: concept to reality in crop improvement. Planta. 239: 543–564.
Scott, J.G., K. Michel, L.C. Bartholomay et al. 2013. Towards the elements of successful insect RNAi. J Insect Physiol. 59: 1212–1221.
Sehgal, D., E. Autrique, R. Singh et al. 2017. Identification of genomic regions for grain yield and yield stability and their epistatic interactions. Sci Rep. 7: 41578.
Shih, P.M. 2015. Photosynthesis and early Earth. Curr Biol. 25: R855–R859.
Shmakov, S., O.O. Abudayyeh, K.S. Makarova et al. 2015. Discovery and functional characterization of diverse class 2 CRISPR-Cas systems. Mol Cell. 60: 385–397.
Sil, B. and S. Jha. 2014. Plants: The future pharmaceutical factory. Am J Pl Sci. 5: 319–327.
Smith, A.E. and D.M. Secoy. 1976. A compendium of inorganic substances used in European pest control before 1850. Agri Food Chem. 24: 1180.
Smith, C.M. and S.I. Clement. 2012. Molecular bases of plant resistance to arthropods. Rev Entomol. 57: 309–328.
Stoeckli, S., K. Mody, C. Gessler et al. 2008. QTL analysis for aphid resistance and growth traits in apple. Tree Genet Gen. 4: 833–847.
Streatfield, S.J. 2007. Approaches to achieve high-level heterologous protein production in plants. Plant Biotechnol J. 5: 2–15.
Swiech, L., M. Heidenreich, A. Banerjee et al. 2015. *In vivo* interrogation of gene function in the mammalian brain using CRISPR-Cas9. Nat Biotechnol. 33: 102–106.
Tabashnik, B.E., T. Brévault and Y. Carriére. 2013. Insect resistance to Bt crops: Lessons from the first billion acres. Nat Biotechnol. 31: 510–521.
Tao, Z., A. Hardy, J. Drenth et al. 2003. Identifications of two different mechanisms for sorghum midge resistance through QTL mapping. Theor Appl Genet. 107: 116–122.
Taylor, E.L., A.G. Holley and M. Kirk. 2007. Pesticide Development, a brief look at the history. Southern Regional Extension Forestry, SREF-FM010. Extension Service publication series. Editor, William G. Hubbard, Southern Regional Extension Forester, 4-402 Forest Resources Building, The University of Georgia, Athens, GA 30602.

Terenius, O., A. Papanicolaou, J.S. Garbutt et al. 2011. RNA interference in Lepidoptera: An overview of successful and unsuccessful studies and implications for experimental design. J Insect Physiol. 57: 231–245.

Thakur, N., J.K. Mundey and S.K. Upadhyay. 2016. RNAi—Implications in entomological research and pest control. *In*: Dr. Ibrokhim Y. Abdurakhmonov (ed.). RNA Interference. InTech.

The Golden Rice Project (last accessed 02.12.17.): http://www.goldenrice.org/.

Timmons, L. and A. Fire. 1998. Specific interference by ingested dsRNA. Nature. 395: 854.

Turner, C.T., M.W. Davy, R.M. MacDiarmid et al. 2006. RNA interference in the light brown apple moth, *Epiphyas postvittana* (Walker) induced by double-stranded RNA feeding. Insect Mol Biol. 15: 383–391.

Unniyampurath, U., R. Pilankatta and M.N. Krishnan. 2016. RNA interference in the age of CRISPR: will CRISPR interfere with RNAi? Int J Mol Sci. 17: 291.

Van Eenennaam, A.L. and A.E. Young. 2014. Prevalence and impacts of genetically engineered feedstuffs on livestock population. J Anim Sci. 92: 4255–4278.

Vianna, G.R., N.B. Cunha, A.M. Murad et al. 2011. Soybeans as bioreactors for biopharmaceuticals and industrial proteins. Genet Mol Res. 10: 1733–1752.

Waltz, E. 2016. Gene-edited CRISPR mushroom escapes US regulation. Nature. 532: 293.

Wang, H., M.L. Russa and L.S. Qi. 2016. CRISPR/Cas9 in genome editing and beyond. Annu Rev Biochem. 85: 227–264.

Wang, X.H., R. Aliyari, W.X. Li et al. 2006. RNA interference directs innate immunity against viruses in adult *Drosophila*. Science. 312: 452–454.

Wang, Y., H. Zhang, H. Li et al. 2011. Second-generation sequencing supply an effective way to screen RNAi targets in large scale for potential application in pest insect control. PLoS ONE. 6: e18644.

Ware, G.W. and D.M. Whitacre. 2004. History of Pesticides, in The Pesticide Book, 6th edition, MeisterPro Information Resources, Willoughby, OH.

Whitten, M.M.A., P.D. Facey, R. Del Sol et al. 2016. Symbiont-mediated RNA interference in insects. Proc R Soc B. 283: 20160042.

Whyard, S., A.D. Singh and S. Wong. 2009. Ingested double-stranded RNAs can act as species-specific insecticides. Insect Biochem Mol Biol. 39: 824–832.

Whyard, S., C.N.G. Erdelyan, A.L. Partridge et al. 2015. Silencing the buzz: a new approach to population suppression of mosquitoes by feeding larvae double-stranded RNAs. Parasites Vect. 8: 96.

Wiedenheft, B., S.H. Sternberg and J.A. Doudna. 2012. RNA-guided genetic silencing systems in bacteria and archaea. Nature. 482: 331–338.

Wilson, R.C. and J.A. Doudna. 2013. Molecular mechanisms of RNA interference. Annu Rev Biophys. 42: 217–239.

Wise, T.A. 2013a. Can We Feed the World in 2050? A Scoping Paper to Assess the Evidence. Global Development and Environment Institute at Tufts University, Medford MA 02155, USA. http://http://www.ase.tufts.edu/gdae/policy_research/FeedWorld2050.html.

Wise, T.A., K. Sundell and M. Brill. 2013b. Rising to the challenge: changing course to feed the world in 2050. Action Aid USA Report, Oct. 2013. Global Development and Environment Institute at Tufts University, Medford MA 02155, USA. http://www.ase.tufts.edu/gdae/Pubs/rp/ActionAid_rising_to_challenge.pdf.

Xie, K. and Y. Yang. 2013. RNA-guided genome editing in plants using a CRISPR-Cas system. Mol Plant. 6: 1975–1983.

Xu, L., B. Zeng, J.E. Noland et al. 2015. The coming of RNA-based pest controls. J Plant Protection. 42: 673–690.

Yadav, B.C., K. Veluthambi and K. Subramaniam. 2006. Host-generated double stranded RNA induces RNAi in plant-parasitic nematodes and protects the host from infection. Mol Biochem Parasitol. 148: 219–222.

Yencho, G.C., M.B. Cohen and P.F. Byrne. 2000. Applications of tagging and mapping insect resistant loci in plants. Annu Rev Entomol. 45: 393–422.

Younis, A., M.I. Siddique, C.K. Kim et al. 2014. RNA interference (RNAi) induced gene silencing: A promising approach of Hi-Tech plant breeding. Int J biol Sci. 10: 1150–1158.

Zha, W., X. Peng, R. Chen et al. 2011. Knockdown of midgut genes by dsRNA-transgenic plant-mediated RNA interference in the Hemipteran insect *Nilaparvata lugens*. PLos ONE. 6: e20504.

Zhang, Y., C. Yin, T. Zhang et al. 2015. CRISPR/gRNA-directed synergistic activation mediator (SAM) induces specific, persistent and robust reactivation of the HIV-1 latent reservoirs. Sci Rep. 5: 16277.

Zhang, J., S.A. Khan, C. Hasse et al. 2015. Full crop protection from an insect pest by expression of long double-stranded RNAs in plastids. Science. 347: 991–994.

Zotti, M. and G. Smagghe. 2015. RNAi technology for insect management and protection of beneficial insects from diseases: lessons, challenges and risk assessments. Neotrop Entomol. 44: 197–213.

Index